Navigating MathLand

Navigating MathLand

How Parents Can Help Their Kids Through the Maze

Linda Kasal Fusco

ROWMAN & LITTLEFIELD
Lanham • Boulder • New York • London

Published by Rowman & Littlefield
A wholly owned subsidiary of The Rowman & Littlefield Publishing Group, Inc.
4501 Forbes Boulevard, Suite 200, Lanham, Maryland 20706
www.rowman.com

Unit A, Whitacre Mews, 26-34 Stannary Street, London SE11 4AB

Copyright © 2017 by Linda Kasal Fusco

All rights reserved. No part of this book may be reproduced in any form or by any electronic or mechanical means, including information storage and retrieval systems, without written permission from the publisher, except by a reviewer who may quote passages in a review.

British Library Cataloguing in Publication Information Available

Library of Congress Cataloging-in-Publication Data

Names: Fusco, Linda Kasal.
Title: Navigating MathLand : how parents can help their kids through the maze / Linda Kasal Fusco.
Other titles: Navigating math land
Description: Lanham : Rowman & Littlefield, [2017] | Includes bibliographical references.
Identifiers: LCCN 2017012696 (print) | LCCN 2017016759 (ebook) | ISBN 9781475824551 (Electronic) | ISBN 9781475824537 (cloth : alk. paper) | ISBN 9781475824544 (pbk. : alk. paper)
Subjects: LCSH: Mathematics--Study and teaching--Parent participation. | Education--Parent participation. | Mathematics--Study and teaching--United States.
Classification: LCC QA135.6 (ebook) | LCC QA135.6 .F86 2017 (print) | DDC 510.71/273--dc23
LC record available at https://lccn.loc.gov/2017012696

∞ ™ The paper used in this publication meets the minimum requirements of American National Standard for Information Sciences Permanence of Paper for Printed Library Materials, ANSI/NISO Z39.48-1992.

Printed in the United States of America

Dedicated to my daughter, Melissa; my loving husband, Ben; and in memoriam of my parents, Gertrude and Ludwig Kasal, and my brother, Dr. Charles Anton Kasal.

Table of Contents

Preface		ix
Introduction: Is Your Child Engaged in Learning Mathematics?		xi
1	Preferences: How Does Your Child Prefer to Learn Mathematics?	1
2	Connects and Disconnects of the School Mathematics Program: What Are the Hurdles Your Child Will Experience in Learning Math in School?	21
3	Teacher: Who Is the Most Influential Person in Teaching Your Child Mathematics?	49
4	American Education System: The Nineteenth-Century Factory Model in the Twenty-First Century: Why Do School Systems Continue to Visit *MathLand* but Never Stay?	69
5	Understanding the Implications of Common Core Standards-Based Programs Accepted in American Culture: How Will Your Child's Math Education Be Influenced by the Common Core Mathematics Standards?	95
6	Next Steps: What Can You Do to Improve Your Child's Interest in Learning Mathematics?	123

Preface

Before we begin, this is the Golden Rule of *Navigating MathLand*: **No one** is to blame for a child not loving how to learn mathematics. Not you, your child, not the school, teacher, union, school system, government, or the "Common Core." Common Core or no Common Core, there is the overarching responsibility of parents, teachers, and schools to provide an engaging math education for all students. However, there are "perfect storms," like Sandy and Katrina, that will impact your child's joy of learning math, too numerous to be identified in the preface.

Once we embrace the blame game, no one wins and we waste time. The intent of this book is to circumvent the blame game by learning to be proactive, rather than reactive, in your child's math education. To do this, you will need to learn to look at the American education system through a different lens and become aware of your child's engagement in learning math. You will also learn strategies to help navigate the thirteen-year voyage your child will take in learning mathematics, and how to avoid potential storms that create hurdles in your child's education.

Navigating MathLand was inspired by the author's work of "teaching in the trenches" and her doctoral research, as well as the teacher training she implements as a consultant. Experiences from her forty-seven-year journey as an educator have honed the tools needed to navigate the system. She majored in mathematics; taught math and science; worked as an administrator (department chair, supervisor of mathematics teachers); and became a doctoral researcher studying how secondary mathematics teachers transition into their practice.

As a consultant she works with teachers, K–12, to develop a math curriculum with and without using the Common Core Mathematics Standards. Her experiences have spanned urban, suburban, and rural districts in New York State. She has taught all ability levels, from special needs students in a self-contained classroom to AP honors students. Her experiences have made her aware of the educational hurdles that deter learning, especially in learning mathematics. Through her research and practice she realized the need to write *Navigating MathLand*—to help parents assist their child in learning mathematics.

Final rule: You may want to begin immediately. Even if you are frustrated with your child's struggles with math, please do not do your child's math homework or "help" them problem solve. What you can do is make sure that your child's addition, subtraction, multiplication, and division facts are mastered. Moreover, you need to monitor your child's acceptance of mathematics and help your child to see mathematics as an integral part in his/her learning.

Introduction

Is Your Child Engaged in Learning Mathematics?

As a parent, do any of the following scenarios ring true from your experiences with your child's math education?

IT'S ELEMENTARY

It's Columbus Day weekend and your child, having entered fourth grade, has math homework to do. You look over the homework sheet, and the questions make no sense to you. The worksheet requires your child to show place value using disks to visualize digits grouped in periods. The homework does not look like the homework you were given by your fourth grade math teacher. At this point, you and your child are both not liking math, arguing, and breaking apart in anger rather than bonding.

The educational jargon defines your child's scorn for math as "not being engaged." You are not alone. According to the Associated Press, an AP-AOL poll of 1,000 people found that at least four out of ten people hated math in school, a dislike that may have been rooted in their elementary school experience.

Last year was a different story. Your child was engaged, did "well," "liked" the teacher, and perhaps did not complain about learning math. Last year's homework was doable for you. However, you learned this past summer that your child scored a level 2 on the third grade state math assessment. That is level 2 out of 4, indicating a failure to "pass" the state mandated test. Level ranges may be categorized by percentages as follows: level 1 (0–54 percent), level 2 (55–64 percent), level 3 (65–86 percent), and level 4 (87–100 percent).

The principal of the school explained at a parents' meeting that the 2013 test, based on the new Common Core State Standards for Mathematics, was more difficult; and lower scores were expected. That, however, does not help you understand why the enjoyable year your child experienced in third grade was not the year for learning mathematics. You ask yourself how some students were able to pass the state assessment while your child was left behind. And now, this year you can't help your child with the homework. Where do you begin to patch the gap and get your child back on board?

The worst part is your child is frustrated and angry, and you are frustrated that you can't help your child. You rationalize that you were a mediocre math student, and perhaps the "apple may not have fallen far from the tree." You are haunted by the desire to want math education to be different for your child, who is already behind and has gaps in his or her understanding of mathematics.

To add to your frustration, on Parents' Night, your child's teacher announces that next spring your child is scheduled to take the grade 4 Common Core State Standards for Mathematics assessment. You ask yourself, "Isn't the purpose of school to attract the student's attention to the subject and to engage the student to persevere and be eager to learn math?"

This introductory scenario summarizes an actual situation that took place in 2013. Maximo was a fourth grader at that time; his parents are John and Vilma. Both parents pursued professional practices as secondary education teachers in the public school system. As readers proceed through *Navigating MathLand*, they will learn how Maximo progressed with his math education through grade 6 in June 2016.

IN THE MIDDLE

This may be a year that your child enters sixth grade and will be exposed to the middle school educational setting, where content areas are departmentalized into math, science, social studies, and English language arts. Content areas that were connected in elementary school curricula are now pulled apart. Even if your student's grade 5 math and science classes were departmentalized, the class remained together as a whole, and the content areas were connected by two elementary teachers.

You attend Parents' Night, where the school principal wails on about the "middle school" philosophy. He or she explains that the middle school experience prepares students for high school academically and socially. To make academics successful for all students, there are teams of teachers that work together to follow your child, making sure that there is continuity in your child's day and validating your child's engagement in learning challenging curricula.

It sounds exciting. Perhaps the principal discusses how the school addresses an "integrated" curriculum that links social studies, science, and English language arts. Somehow, math has been disconnected from the integrated curriculum. The middle school math teacher presents the new Common Core learning standards curriculum that is based on applying math concepts and skills to solve "real-world problems."

However, if math is taught apart from the other subjects, how does the Common Core math curriculum address real-world problems encountered in the science curriculum for grade 6? Keeping math as an isolated subject doesn't make sense, it doesn't connect.

You may be fortunate to live in a school community where the families are in a range of middle- to upper-level socioeconomic status. In New York State, you are able to access your district's middle school results for the 2013 state mathematics and ELA (English language arts) Common Core assessments for grades 5–8. At each grade level you see that, at most, 60 percent of the students were able to score a level 3 or level 4, which is considered passing.

NYC charter school students have achieved higher scores. You question how effective your child's math program has been in helping students pass the state assessments.

When you inquire about which math textbook is being used, the answers given by teachers may vary from "no text," to "resources (workbooks, worksheets)," to "a standards-based program that is aligned with the Common Core." Oftentimes, your child comes home with worksheets with problems that do not look like the math problems you solved when you attended middle school. In fact, the math text that your child brought home last year might have been replaced by a set of worksheets from a state "Common Core" website. When you go to help your child, there is no text that you can reference.

In middle school, math is your child's least-liked subject. Even if your child is getting As or Bs, he or she is not enthusiastic about learning more about math. At the end of the year it will be determined whether or not your child will be able to handle the honors math program in grade 7 (a decision usually preordained by the local entrance exam created by the middle school math department).

Not to worry, if your child does not pass the entrance exam, they can go to summer school and take an enrichment course that will certify them to rub elbows with the elite "wunderkind" in the fall. Giving up a summer for math is not a bell ringer for the prepubescent middle school student, especially if the honors selection process labels the student an "also ran." Not making the cut for the "advanced" class does not make for an even playing field or further enhance your child's desire to learn math.

It may be that the principal of the middle school has not shared each student's score on the past spring state exam. Grade 6 teachers may not know who in their classes achieved only a level 1 or 2 on the grade 5 spring state assessment, making it difficult for them to know how to differentiate instruction for your child. If your child scored a level 4, they may have to sit through a boring review of fifth grade math. On the other hand, the child who scores a level 1 or 2 has gaps that might need to be addressed before they can be introduced to the grade 6 math curriculum.

AT HIGH SCHOOL

Your child is either one of the chosen accelerated students and will be taking geometry as a freshman or, if not selected, they will be taking Algebra I, perhaps with ninth and/or tenth graders that are "middle of the road" (that is, children already branded as second-class citizens because they did not make the cut for the middle school accelerated math program of Algebra I in grade 8). Your child might have scored a level 3 on the state test, which gives him or her the OK to take Algebra I in grade 9.

The math teacher may not expect high performances from non-honors students, and may believe that it is only the honors students who can handle the rigors of algebra, geometry, and trigonometry. It is common to find math teachers who believe the Common Core Algebra I math curriculum is too challenging for even the honors students. These opinions synergize the already-present stigma that non-honors students are mediocre and not capable of learning a more rigorous math curriculum.

It is not prudent to place blame on a teacher who is immersed in a culture where level 3 students score threes and level 4 students score fours; and likewise, for students who score at levels 1 and 2. Now there is new responsibility for teachers to raise the scores of students who get 1s to 2s, and 2s to 3s. That responsibility is tied to the teacher's yearly evaluation. The stress is on the teacher to improve the "test" scores. The instructional decisions made by teachers are often basically to "teach to the test," rather than to the Common Core math curriculum.

At either end (pre-K and college/career) of your child's math journey, there are markers that impact achievement. If your child is identified as a special needs child, then the journey is complex. Parents look to the future of their child's learning, wondering what educational tools they will need to get into college and/or to select a career. Parents want their children to be passionate about their work; it helps if they know how to best engage them in learning.

THE FIVE CRITERIA FOR ENGAGEMENT

By the time your child reaches high school, their learning level is preordained. Generally, students in non-honors math classes consider themselves not worthy of engaging in the rigorous study of math. This is true for up to 90 percent of the student body.

The big question is, "What does engagement in learning math look like?" The Oxford English dictionary (http://www.oxforddictionaries.com/definition/english/engage) defines the verb "engage":

> [*with object*] Occupy or attract (someone's interest or attention)

Example: "The teacher attempted to engage the class in the mathematics lesson." The definition goes further to include:

> (engage with) [*no object*] Participate or become involved in

Example: "Students engage in exciting math tasks." In fact, the New York State Education Department (NYSED) website for the Common Core Learning Standards (for mathematics and ELA) is aptly named "EngageNY."

What does it mean to engage a student in math? How do you know your student is engaged in learning mathematics? You may want to be more concerned about the "self esteem" trap set by teachers who dole out As and Bs for effort, doing homework, and class participation rather than how the student scores on math exams. It's self-efficacy that is important. That is, being honest with a student. It should be clear where a student is strong and where he or she is weak; then work with that student to improve the weaknesses.

Think back to some task that you knew you could not do (were weak in)—like learning how to ride a bicycle. You first were given training wheels. After getting a "feel" for riding the bike (teetering back and forth), the training wheels were removed and your parent or some close adult held you steady while you peddled. You started to feel what it was like to balance without the training wheels. Then you were released on your own and you could balance on the bike. You learned to ride a two-wheeler, and today you can still get on a two-wheeler and ride. That sense of accomplishment is what a student who is learning math concepts and skills needs to feel. Tasks that used to be issues can now be done independently on their own.

Let's review the learning process. You did fall off the bike, and you were allowed to make mistakes. You admitted to yourself that you made mistakes riding and you self-corrected the mistakes. You collaborated with your parents and friends for the best way to ride. There was no timeframe for learning; you did it at your own pace.

Like getting a driver's license, we "engage" in learning how to drive. All people of all abilities can learn how to drive. They are motivated to learn; it's a rite of passage in life. Just as driving a car or riding a bike is applicable to our lives, so is learning math. The key is "engagement" in the subject.

Engagement may or may not be a characteristic found in your child's math class. However, understanding how your child prefers to learn math is key to how you can help your child embrace math instruction, with or without the Common Core. But there are many factors besides these preferences that will impact your child's thirteen-year journey.

Not only do we want to know what engagement looks like, we also want to know how to ensure that our children are productively engaged. In the elementary scenario Maximo liked third grade and liked the teacher, but was not well prepared by the teacher to take the grade 3 Common Core math state exam. The school year was over and it was too late for Maximo's parents to intervene. The parents were left without any feedback on what constituted Maximo's strengths and weaknesses in math concepts and skills.

According to "Edutopia: Engaging Students in Math," a report on student engagement from the George Lucas Educational Foundation, there are five engagement criteria that foster the optimum environment for engaging students in learning. The following principles assure students a learning environment in which to meet success:

1. Students are allowed to make more mistakes, and then encouraged to work through their mistakes, and learn from their mistakes.
2. There is a classroom format to support the struggle of those students who need time processing math skills and concepts. It is OK for students to struggle and learn how to persevere and become self-motivated learners.
3. There are also ample opportunities to let students "teach" by having them argue their solution with their peers. Students are encouraged to vocalize a problem and share their difficulties with the class.
4. The curriculum provides gray area questions, not just questions that are right or wrong. An instruction such as "find the number of rectangles that can be generated by the factors of the numbers 2 through 36" has many correct answers.
5. The teacher may personalize the questions. You would be surprised how inserting student names in problems gives a sense of ownership to the problems. (https://www.edutopia.org/blog/engaging-students-in-math-jose-vilson)

THE IMPORTANCE OF KNOWING A CHILD'S LEARNING PREFERENCES

Not only is the learning environment essential for engaging students, an awareness of student learning styles in creating lessons that motivate students is also important. *Navigating MathLand* will provide you with a new lens through which to view how your child learns math, a lens that is essential to engaging students in learning mathematics but one that is rarely applied in classroom instruction. The focus of traditional math lessons is content-driven, leaving the delivery process of the concepts and skills needed to understand math on the back burner. Common Core provides engaging math instruction for all students.

Your child may have been perfect this year and had a flawless experience in math courses. Next year may be a different experience, like the third grader who loved school but then lost that passion for learning in fourth grade. Wherever your child lands on the spectrum of "lost in

MathLand" to "solving genius level *MathLand* problems," it is very helpful for students to understand how best they learn mathematics.

PREVIEW OF *NAVIGATING MATHLAND*

In chapter 1 ("Preferences—How Does Your Child Prefer to Learn Mathematics?") you will have the opportunity to investigate your own math learning style, and to discover how knowing the math learning style profiles of each student can help teachers and parents to foster students' engagement in learning math. What is a math learning style? Here's a preview—Does your child (elementary, middle, or high school) fit the following scenarios (styles)? Is he or she a student who:

a. likes the teacher who shows the step-by-step procedure to solving math problems and asks "What? What are the facts?"
b. is inspired by a teacher who makes math interesting and challenging and asks "Why? Do I understand why the step-by-step procedure works?"
c. enjoys the creativity of mathematics and asks "What if? What if there was another way to solve the problem? What if we changed some of the numbers?"
d. would like to discuss math but is asked to keep quiet in class and exclaims, "So what! So, what is this math really useful for?"

These four preferences are linked to students' math learning styles. Parents will learn that students exhibit all four styles, with one style being dominant.

There are students who learn math step by step, through procedures—the "Mastery" style. Students who like math problems that ask them to explain, prove, or take a position exhibit the "Understanding" style. Creative students exhibit a "Self-Expressive" learning style and like to use their imagination to explore mathematical ideas that allow them to think "outside the box." And students who exhibit an "Interpersonal" learning style learn mathematics best through dialogue, collaboration, and cooperative learning.

You're told your child is brilliant but not motivated to learn mathematics; or your child has aced the algebra exam but can't make sense of geometry—can't understand proofs. Perhaps your child was a C student in algebra but is excelling in geometry. Your child may have been a straight A student in mathematics until he or she reached calculus and then struggled with college-level math concepts.

The bottom line is that a student has thirteen years of education in which to become math proficient. The path to learning mathematics is not consistent. Students "drop out" of being curious and self-reliant as problem solvers. By the end of the thirteen years, 90 percent of students, K–12, do not get into honors math courses, leading students to believe they are not "good at" math. Many cannot attain career goals because they believe they are inept in math.

The recurring question is, "What is the importance of math in our lives?" Perhaps a student who knows how she or he best learns mathematics (their math learning preference profile) early on may avoid missteps in their thirteen-year journey though school. Chapter 1 will guide the reader and demonstrate how understanding math learning styles is essential to helping students and provides vital information to teachers for developing engaging lessons.

You want to be proactive in navigating the education system. Starting with chapter 1, you will learn strategies that will help you guide your child through his or her math education. There are instructional methods based on the four math learning styles that teachers can use to engage students in mathematics. These preferences will be discussed in depth in chapter 1.

CONNECTING MATH PROGRAMS WITH STUDENT LEARNING PREFERENCES

Implementing a standards-based math program, like the Common Core curriculum, into a school district is a Herculean endeavor. School districts have been attempting to implement rigorous standards-based programs for the past three decades. Inevitably, "starts and fits" arise within these programs, leaving a bumpy learning landscape in which students must navigate. Students often experience academic hurdles as a result of programs that have been started and then discontinued.

Chapter 2 ("Connects and Disconnects of the School Mathematics Program—What Are the Hurdles Your Child Will Experience in Learning Mathematics in School?") will lay out the learning issues born out of poor program implementation. Not only on the district level are there issues with programs but edicts from the state and federal government also create havoc with classroom instruction. The rollout of the New York State P–12 Common Core Learning Standards for Mathematics was the fourth set of state standards to be instituted in thirteen years.

Knowing how to become a proficient math student is a primary hurdle. Proficiency that is based on making mathematics meaningful to a student's life is as important as learning how to drive. There are other hurdles, such as students being grouped into classes by age rather than by their ability to learn math, school schedules and summer breaks, and lost years due to misplacement in math classes or poor teaching, to name a few. Another hurdle for students and teachers is that there is often not a clear understanding of what the study of mathematics is.

What is mathematics? We all have a pretty clear idea about what we learn in biology, American history, English, music, and art. Students, for the most part, easily engage in subjects that provide clear, concrete curricula that apply to their lives. But how do we engage students in a subject that is intangible and abstract? Look at your child's math text and try finding a definition of mathematics. Conducting a philosophical discussion about "what mathematics is" is rare for mathematics teachers, and rarer for their students.

At the beginning of each school year, some students are lucky enough to get a copy of a math curriculum that covers the topics: Number Systems (the real numbers); Expressions; Solving Equations; Graphing Equations; Introduction to Trigonometry; Polynomials; and Geometry. Students may or may not bring home a textbook (because the teacher may create worksheets), but nowhere in any resources is there an introduction as to what constitutes a true study of mathematics.

Science teachers can define their subject. Biology is the study of living systems. In fact, science texts usually use the introductory chapter to define the science germane to the course (that is, chemistry, earth science, forensics, physics). Music and art are clearly defined, as is English and history. Mathematics is never defined for the student.

Compounding the lack of clarity of the subject of mathematics is the fact that students usually lack an awareness of their preferences as to how they learn mathematics. Student preferences for learning mathematics are rarely studied in college methods classes attended by preservice teachers or in professional development workshops conducted for practicing teachers (research has shown that students have a dominant style when learning mathematics). Students tend to be engaged in learning math when the instruction is geared toward their preferred learning styles.

Exciting math education can take place with or without the Common Core. Past standards-based programs, such as *MathLand*, were developed based on "best teaching practices." A best practice is a research-based instructional method that a teacher uses to engage students in learning. Examples of best practices were the five aforementioned criteria a teacher can use to create an engaging learning environment (such as providing recognition and immediate feedback).

COMMUNICATION ABOUT YOUR CHILD'S MATH PROGRAM

MathLand was one of the first standards-based math programs developed in 1989 based on teaching math as a process/concept-driven program as compared to the traditional content/skills curricula. *Navigating MathLand* is about the return to *MathLand* standards-based programs like the Common Core. The curriculum was developed in the late 1980s.

What follows is a letter introducing the *MathLand* program to the parents of sixth graders sent by the math teacher that explains how their children will learn math for the school year. The letter communicates some of the best practices students will be experiencing in math class. As a parent, have you ever received a letter like the one below?

> Dear Family,
>
> This year your child will be learning mathematics the *MathLand* way — by doing a variety of rich mathematical projects. You can expect a difference this year, both in the way your child is being taught mathematics and in your child's attitude and sense of excitement about mathematics.
>
> The *MathLand* program is organized around exciting weekly projects that will involve your child in thinking, writing, talking, and doing mathematics. You won't be seeing worksheets this year. Instead, your child will generate recordings and reports on blank paper. Your child will still learn basic computation skills in math class. These skills are important tools your child needs to solve many of the higher order thinking problems we will be doing this year. At reporting time, when we will look through your child's portfolio of work, you will gain many insights about your child's mathematics thinking — insights that would not have been possible from looking at traditional fill-in-the-blank worksheets.

Your child will work with a partner on most projects. By working in pairs, children learn to interact successfully with others, and better thinking results when children (or adults) have a chance to exchange ideas. There will be times when the class explores an idea together, creating a large wall chart to show our work. Come to our class and see our *MathLand* bulletin board!

Because your child's learning need not be confined to school, your child will receive a Student Letter each week telling what we have been doing in *MathLand* and offering a homework suggestion for your child to do, sometimes with a family member. Sometimes your child will be asked to send the results of an activity back to school. Save and repeat activities that you think are especially fun or beneficial for your child.

There are some ways in which you can help. If you have access to paper that we can recycle by using the back side, please send it in. We also can use colored paper, card stock, and cardboard. In addition we may ask you to save and send to school such things as empty boxes, cartons, cans, or plastic jars for some of our projects.

It's going to be a great year. We are looking forward to sharing it with you.

Sincerely,

Did your child ever receive a homework assignment like this? Could you help your student with the following grade 6 homework assignment?

Dear Student,

This week you have been making up number rules on your own and plotting them on coordinate grids to see the kind of lines they form. Another common use of the grid systems is on maps.

Find a map of the city or country in which you live.

1. Find the location of your home on the map. What are the coordinates?
2. Find the location of some of your city's or county's "hot spots" (fun places, restaurants, parks, and other favorite spots) that you think visitors to your city or country might like to see. List the coordinates of these places.
3. Bring the map you use to class along with any other maps you can find that use the coordinate grid system.

(*MathLand Journeys Through Mathematics: Reproducibles Grade 6 (Family Letters, Teaching Resources)*, by Linda Charles, et al., Creative Publications (1995))

As a parent, do you find yourself engaged by these letters? Standards-based programs should be an essential part of teacher training. However, standards-based training is not a big focus for teachers transitioning from training into practice. Having a bank of projects provided for the new teacher, as a program like *MathLand* does, affords more opportunities to implement productive and creative ways to engage students.

THE EDUCATION PACKAGE:
MATH TEACHERS AND THE AMERICAN EDUCATION SYSTEM

The next "hot issue" is—the teacher. What is the teacher's role in providing engaging lessons for students? Keep in mind that we will not invoke the "blame game" or look at teachers as the "victims" of whomever (boards of education, administrators, parent pressure). In chapter 3 ("Teacher—Who Is the Most Influential Person in Teaching Your Child Mathematics?"), you will learn what to look for in instruction that engages students in learning mathematics.

With or without the Common Core, the task is to get the students to drive the *MathLand* road through engaging teaching. The pivotal person is the teacher. In chapter 3 you will learn how teachers' beliefs, which include how mathematics is defined and taught, impact their instructional decisions. Teachers need to be aware of how their philosophy of mathematics directs their instructional interpretation of the curriculum, and how they embrace standards-based math programs.

For example, just as there are learning styles for math, there are complementary teacher styles that provide the environment in which your child learns best. Also, be aware that in the real world, we encounter scenarios that don't fit how we learn. Especially with regard to mathematics, a student needs to be aware of how to navigate in an environment that may not be ideally suited to his or her learning style and what to do when they find themselves in a class where the teaching method is not conducive to the way they learn.

Not all the learning issues are a result of teaching and school curricula. There are barriers produced by an archaic American education system that restrict the change necessary to implement appropriate twenty-first-century instruction. It is important to learn about the barriers, such as the 45-minute, nine-period "daily" schedule and the "same grade, same age" conundrum. Is everybody in your workplace the same age?

In chapter 4 ("American Education System—The Nineteenth-Century Factory Model in the Twenty-First Century: Why Do School Systems Continue to Visit *MathLand* but Never Stay?"), the barriers will be identified, accompanied by a brief history of the evolution of the Common Core through the centuries. Here's a peek into chapter 4.

In 1970, junior high schools typically addressed grades 7 and 8. Some junior high schools had three grade levels and included grade 9. Before the idea of junior high took place in the 1950s, the elementary schools ran K–8. It seemed that the seminal rationale for developing a "middle level school" for children was to separate the pubescent junior high students from the elementary students. Somehow, educators of that era believed that the persona of an eighth grader had a negative influence on a second grader.

Gone was the one-room schoolhouse concept where the older students could support the learning of the younger ones. This "middle school" move also extracted the teachers who were content specialists (math, science, etc.) from the elementary teachers who were generalists and whole-child oriented. Besides, the bang for the buck was better, as we could warehouse more students in a junior high (which was modeled after the senior high school's nine-period day, 45-minute class schedule developed at the end of the nineteenth century).

SHIFTS IN MATH CURRICULUM AND INSTRUCTION

Currently, the middle school culture is shifting, as the Common Core has placed high academic demands on students and teachers. The math instruction has shifted from the traditional procedural "drill and kill" to having lessons to develop understanding. For example, the CPA (Concrete, Pictorial, Abstract) instructional strategy is the basis of the math lesson.

Formerly, a math lesson was all about how to execute the algorithm. Children learned how to cross out a number and "borrow" 10 in order to subtract a column of numbers. Now, teachers introduce subtraction using a place value chart or place value disks, where students can visually experience borrowing 10. Creating understanding of a math concept and/or skill is built up from concrete experience, with a visual representation, to the algorithm.

CPA is further addressed in chapter 5 ("Understanding the Implications of Common Core Standards-Based Programs Accepted in American Culture—How Will Your Child's Math Education Be Influenced by the Common Core Mathematics Standards?"). The reader will learn how the Common Core curriculum has risen from the math reform ashes (like the Phoenix) to deliver said relevancy as to why learn math. With or without the Common Core, the purpose for learning math needs to be clear to learners. We want students to become math literate, just like they are driving literate.

As a parent, it is important to develop your understanding of what the Common Core means to you. Understanding the difference between a standards-based curriculum and the traditional content-based curriculum will help you avoid the rumor mill and the ill-informed hysteria that results from educational reform. A poor rollout of the Common Core math curriculum, inadequate training of teachers, and a plethora of misunderstandings about the program from all sides (teachers, parents, administrators, etc.) can do more harm than good.

It is also important to understand the evolution of the Common Core so that you can navigate the thirteen-year math education journey of your child. Yes, history does repeat itself and oftentimes students are the spoils of the "math wars" and errant pedagogical experiments, such as middle school.

For the math teachers, the paradigm shift to the Common Core standards-based math instruction has been difficult. On top of the added requirement for teachers to instruct for understanding, the No Child Left Behind Act (2001) required that all students have a rigorous program. More importantly, all teachers must now be state certified in order to teach secondary math programs (including special needs). Special needs teachers were given until 2014 to become certified in mathematics. In 2015, the Every Student Succeeds Act (ESSA) was passed, which created provisions to reach the goal of fully preparing all students for success in college and careers.

IMPACT OF AN INSTRUCTIONAL SHIFT

The rules of engagement in learning have been changed by the national Common Core State Learning Standards for Mathematics. Students coming into middle school only achieving a score of level 1 or level 2 on the grade 5 state math test are now identified as not on grade level, but what does the middle school do with the drove of students who are not proficient in math (that is, those grade 5 students who have not "passed" with a level 3)? If your child has scored a level 1 or 2, chances are there will be a divide in the class, and your child may remain in lower level classes throughout high school.

By high school, your child's ability to learn mathematics has been determined. Middle school students who have advanced in math have been separated from the level 1 and level 2 students; and yes, some level 3 students. The school has already decided which students will succeed in mathematics, leaving no chance for the middle-level student to advance, thus placing them in the mediocre math classes, with perhaps mediocre instruction.

Then comes the realization for high school students that they only have to get through algebra (maybe it will take two years) and some geometry, and maybe trigonometry. By avoiding a rigorous math program, they do not realize they are limiting their chances to score well on SATs and to pass the physical sciences (physics, chemistry, earth science) that require mastery of algebra. At this point, parents already have accepted the fact that their child is "weak," "unmotivated," and "disinterested" in mathematics, and certainly not engaged.

Unfortunately, in the United States, we are talking about the bulk of the students. Even the advanced students find math tedious and a means to an end—that is getting high scores on SATs and getting into a "good" college. And so the thirteen-year journey that started out tabula rasa for your child has morphed into a no-win situation that could possibly limit your child's post–high school experiences (college and/or competing in the twenty-first-century "global economy" job market).

The Common Core State Standards for Mathematics provide a template for forty-six states that have adopted the Common Core to develop lesson guides for teachers with challenging problems on how to engage students in math discussion and to work through challenging problems (for example, EngageNY math modules). Any parent from any state can review the EngageNY math module resources. Chapter 5 will provide a more in-depth explanation of such Common Core resources.

PARENTS: BE PROACTIVE IN HELPING YOUR CHILD NAVIGATE LEARNING MATHEMATICS

Parents need to learn about the math program selected (by the district), or find out if a program has been selected at all, to ensure that "best practices" are being used in math instruction. Remember that the most influential person that your child spends time with during the day is the math teacher. The rule of thumb is that your child will have the same teacher (barring unforeseen circumstances) for the entire year. The relationship that your child forms with that teacher will make or break the learning of math skills and concepts. Getting a tutor is not the total answer.

As a parent, you need to understand how you can help your child enjoy learning and applying mathematics. Do not just stand by but be proactive in guiding your child through *MathLand*. You will find out how to help your child through the haze and maze of learning math, with or without the Common Core, in chapter 6 ("Next Steps—What Can You Do to Improve Your Child's Interest in Learning Mathematics?").

Learning math can be equated with the importance students place on obtaining a driver's license. Students from all different "walks" of education learn how to drive and are able to tackle the skill of driving without incident. There are physics concepts students need to know about how a car navigates a turn, the effects of applying the brakes in time, and how to accelerate the car to a posted speed limit. Students need to be knowledgeable about speed, acceleration, and stopping distances as well as the language of the highway and the rules of driving.

All states share to some extent a Common Core driving curriculum—a curriculum based on the goal of keeping drivers safe on the roads. For all students (underachievers to brainiacs), driving is important. Driving is a rite of passage into adulthood. To boot, students develop enthusiasm for learning to drive outside of the formal academic curricula presented in school. To learn how to drive, students must pass a written test and a road test.

The driving assessment is "hands-on" and the test is universal. Once the student earns a driver's license, he or she will be set for life to drive the family car; or they may decide to further their driving education and become licensed to drive other vehicles. There is a general driving license and there are upper-level licenses—for a limo, cab, bus, and so on. These driving skills may be transferred to other forms of transportation such as learning how to pilot a boat or fly a plane.

Parents, or other qualified adults, are usually responsible for guiding children through the process of learning to drive a car. However, the motivation to learn how to drive is rooted in early childhood. As toddlers, all children are thrilled to move in a toy vehicle. Later on, they'll learn to ride their first bike or scooter.

As teenagers, most students are motivated to study for their permit test and practice for their road test. They are engaged in learning to drive. Student drivers learn that they need to pay attention to what is going on while they are driving a car. The content and process of learning how to drive makes sense to them. If they fail the first driving test, they get back on their feet and try, again and again, until they have reached their goal. Students persevere to get their driver's license.

Parents act as academic teachers do, by providing a safe environment in which to help their children learn from driving mistakes. Taking your child to a vacant parking lot to practice is like a teacher providing a safe environment in the classroom where students can make mistakes and learn from their mistakes. There also is time for each new driver to develop driving skills at their own rate, without being labeled as "slow."

Wouldn't it be great if students were engaged in learning math, with the same perseverance as learning to drive a car? Could math be "hands on" interesting and relevant to your child, challenging them with opportunities to redo "tests" until they have mastered the math skills and concepts? Math curricula developed to align with the Common Core State Standards for Mathematics tout engagement of students to master math skills and internalize understanding of math concepts, making "Why learn math?" as relevant a question as "Why learn to drive?"

What does engagement look like? How can you as a parent understand how your child engages in learning math, with or without the Common Core? Remember the name of the New York State Education Department website for the Common Core math curriculum is EngageNY. However, as a parent, you do not have time to wait until the Common Core movement "kicks in" (or not) nor energy to waste blaming the education system, the teachers, or your child.

ONE
Preferences

How Does Your Child Prefer to Learn Mathematics?

Have you ever had a conversation with your child's teacher about how your child prefers to learn math? Or have you discussed how the teacher implements instruction based on his or her understandings of how to "hook" your child into the math lesson? The following scenario happened as reported.

In 2005, a parent requested a meeting with the supervisor of mathematics for the Harrison Central School District (NY) regarding her son's experience in the seventh grade math class. According to the parent, in class her son was achieving grades at his usual A level. He was, however, complaining about having to work in a group.

The supervisor was able to access the child's preferred math learning style from the school database. Her son's profile listed "Understanding" as his dominant style to learn mathematics. "Understanding" style students want to know why the math they learn works. However, students with a preferred "Understanding" learning style may experience difficulty when there is a focus on the social environment of the classroom, such as cooperative problem solving.

The supervisor could easily explain to the parent that *MathScape*, the standards-based math program adopted by the math department at Louis M. Klein Middle School (Harrison, New York), required, at times, that students discuss solutions to math problems in groups. Those students who exhibit an "Understanding" math learning style may express discomfort when having to discuss their answers with others.

The supervisor assured the parent that the *MathScape* program also provided experiences for students to work alone, and provided problems that asked students to explain their solutions in writing. Working alone to find a solution to a challenging problem is an attribute of the style that was more apt to engage a student with an "Understanding" math style learning profile.

Her child was feeling uncomfortable because the class at that time was working on a project as a group. The supervisor explained to the parent that it was important for students to discuss solutions to math problems with others. The group work might have been outside her child's comfort zone. In the real world, a person who has found a solution to a problem needs to be able to communicate the solution to others. It is important that the child recognizes the discomfort, and works on communicating his ideas to others.

The parent was very satisfied to learn how her child preferred to learn math and why, at this time, he was feeling uncomfortable. The supervisor reviewed the remaining three of the child's learning styles and demonstrated to the parent how *MathScape* (a standards-based program) addressed his math learning style profile.

On more than one occasion the supervisor was able to pull up a child's math learning style profile; and the discussion was about how the student prefers to learn the curriculum. Learning profiles are essential in diagnosing and providing a student with an instructional plan. For the teacher, learning profiles provide a baseline for developing math instruction that can engage students in learning math.

However, how students prefer to learn math is rarely on the educational radar. For the amount of time (thirteen years) that students spend in school, three-quarters of their math education journey (grades 3–12) is subject to a math "content" curriculum presented by topic, not the engaging processes needed to learn the content. The curriculum is "covered" whether or not the students internalize the concepts and skills necessary to achieve a passing score (level 3) on the end-of-the-year state assessments.

What indicator does the teacher have to understand student achievement? The state assessment is a normed test that was designed to identify student achievement. The purpose of a test is to diagnose a student's strengths and weaknesses, including any "gaps" in expected standards for that grade level. Feedback from the April state assessments on how students achieved grade level math standards is to be used by the math teachers to differentiate instruction for the students entering the next grade in September.

Administrators often do not have the opportunity to share student data garnered from the assessments (that is, what levels were scored by their new batch of students) with teachers. Teachers often have no idea if their students scored a level 2, 3, or 4; they are not directed to use state data to drive their instruction nor are they shown how to provide engaging review developed for each child's weakness. However, knowing the learning style of a student is essential to teachers in developing engaging lessons regardless of how students performed on the state exam.

DIAGNOSING HOW STUDENTS PREFER TO LEARN MATH

By the time students enter high school, there is a plethora of assessment information for each of them on file in the guidance office. However, all of the diagnostics and whole-child evaluations that were completed in elementary school tend to be buried or lost as a student transitions from elementary to middle school. It becomes more difficult and time-consuming for high school teachers to access student learning histories.

Teachers, as clinicians, should be steeped in knowing about each student's journey in learning math. But there is no time set aside for teachers, especially at the secondary level, to review student files in September. Teachers would need time to review files for 150 new students. However, simply knowing the styles provides a quick way to assess how students prefer to learn math. That is why having a child, as well as his or her teacher, understand their learning preferences will help them when entering a new school year.

It is necessary for a teacher to be able to assess the math concepts and skills that need to be improved in order to prescribe the proper instruction. Knowing how a student prefers to learn math helps the teacher provide engaging lessons. The teacher acts like a medical doctor and creates a learning plan.

Think about a visit to your doctor. You have symptoms of an ailment that need to be identified and remedied. Just suppose the doctor gave you a generic cure for your symptoms, like a generic curriculum, without creating or referring to the profile (medical record) of your health. Prior to being examined by your doctor, you spend time to input your own and your family's present and past health history into a database, your medical file.

The doctor then proceeds to gather more real-time data by running tests on your body fluids and taking your vital signs (blood pressure, etc.). Included in the process is a hands-on physical exam; perhaps the doctor uses "state of the art" medical instruments to round out the database. Hopefully, your physician has kept up with twenty-first-century medical techniques and is knowledgeable in his or her specialty. You share your preferences for treatment options.

All of these steps are done prior to your doctor's diagnosis of your ailment. Finally, the doctor prescribes a regimen to address your ailment based on your tests and examination. Your condition improves or it may be chronic, but you usually know what is wrong. Just as your doctor is a medical specialist, your child's teacher is the math education specialist.

More often than not, on Parents' Night you find out about what topics your child will learn this year in mathematics. But rarely will the teacher diagnose how your child learns and explain how he or she will tailor this year's math curriculum to your child's math learning "diagnosis."

There are archives and data that teachers and parents can access that are helpful in identifying how your child learns. The state assessments are not the only indicators of your child's ability. Elementary schools pride themselves in providing detailed information about how a child learns and how the child scores on standardized tests (reading levels, math ability and skill levels). The elementary files for students include notes from teachers that give some "diagnosis" or profile as to what learning tasks your child has completed.

The information is passed on from teacher to teacher from grades K–5, and the file grows. How your child learns in elementary school is addressed from the whole-child perspective. When a student moves up from elementary to middle school, the information travels to a designated team of educators who decide into which program(s) he or she will be placed.

Based on the elementary profile, your child may be placed according to abilities and skills on a team with perhaps 150 other children. The middle school program may not have either an accelerated math program or a math support program. If your child scored a level 2 on the state exam he or she will be destined for the support program, and perhaps miss being included in a more challenging program.

If a child has not had the opportunity to learn math in his or her preferred style, that doesn't mean he or she can't learn rigorous mathematics. Math skills and content need to be taught in the child's preferred style. Support math programs need to be designed to "scaffold" up to rigor, starting with instruction that engages children in the manner in which they prefer to learn math. Support programs often deliver math support in the same way math has been taught in the classroom. The result is that the student shuts down and does not take advantage of any support.

By high school, the elementary diagnostic information on your child is usually buried in a guidance department file cabinet, along with the middle school report cards that don't matter (because now high school transcripts are the focus). If your child scored at level 2 on the grade 8 state math assessments, he or she may be scheduled for math support in high school. Rarely do students in level 2 become level 3s (or passing). A student who scored a level 2 in grade 8 may struggle with getting through the ninth grade Common Core Algebra I course.

By high school your child may not be engaged in learning mathematics, and may believe that he or she can't learn math. In 2013, even in the highly rated school districts in New York State, only 60 percent of students reached a level 3 on the middle school math assessment. By 2016 there had been a slight increase across the board on the New York State grade 3–8 math assessments.

If there is a question about the ability of students to learn math, middle and high school teachers need to have the opportunity to review the archives and read students' folders to find clues to past "elementary" history in learning math. There is "data" teachers can use to drive their instruction to be engaging and useful in improving student scores on state exams, but most important is getting students to embrace learning.

In this chapter you will understand that knowing how children view themselves as math learners is seminal to engagement in learning math. Knowing students' math learning profiles helps teachers to differentiate instruction for both large (25–35 students) and small (8–24 students) classes. When a teacher can differentiate instruction, there is a better chance to interest students in learning mathematics. Without understanding student preferences, an instructional failure to "hook" or engage students in the math lesson occurs.

There are instruments designed to determine how children prefer to learn in general. In digging into your child's elementary school file you may find specific instruments used for diagnosis. You will find tests that were administered to identify what math skills, concepts, and ability level your child achieved. You might even discover tests to identify preferences—such as what career paths your child might take.

MATH LEARNING STYLE INVENTORY

The *Math Learning Style Inventory (MLSI) for Secondary Students Grades 6–12*, developed by Harvey F. Silver and colleagues, is specifically designed to identify how your child prefers to learn math (Silver, Thomas, and Perini, Thoughtful Education Press, LLC, 2008, www.ThoughtfulClassroom.com). Math learning styles are considered "soft skills" and have not been embraced or integrated by school districts into the district database. The traditional school setting concentrates on content. What is content? Content is knowing multiplication tables, shapes of figures, the division algorithm, and solving an equation with an unknown variable "x."

Soft skills (e.g., communication, teamwork, presentation) are personal attributes that enhance an individual's interactions, job performance, and career prospects. Unlike hard skills, which refer to a person's skill set and ability to perform a certain type of task or activity, soft skills relate to a person's ability to interact effectively with coworkers and customers, and are broadly applicable both in and outside the workplace (https://bemycareercoach.com/soft-skills/list-soft-skills.html).

Just as there are "soft skills" profiles for adults, learning profiles have been developed for children that identify learning preferences. In particular, the *MLSI* will provide you and your child with a snapshot of preferences for learning mathematics. In this chapter you will have a peek as to what questions are asked on the *MLSI* and how the profile is scored.

The goal of this chapter is to give you a sense of what a learning profile is, how it is used, and why it is important. The impetus for a student knowing how they prefer to learn math will be the basis for understanding how the school environment both enhances and impedes engagement in learning mathematics.

The *Math Learning Style Inventory* is implemented in classrooms by school districts, individual teachers, administrators, and consultants. Developed at the beginning of the twenty-first century by Thoughtful Education Press, LLC, this instrument is introduced as follows: "Math is all about problem solving. But not all students and not all mathematicians solve problems in the same way. In fact, even though your textbook might tell you otherwise, there are many different ways to solve math problems."

The *MLSI* is a classroom-tested tool that allows teachers to identify how each of their students prefers to learn mathematics. Knowing students' preferred learning styles helps teachers to differentiate math instruction so that students' needs and styles are addressed in classroom instruction, attending to the range of students from at-risk to gifted who may not be achieving their full potential. Teachers starting their practice may have learned in their teacher training how to use learning styles to design lessons.

Teachers, especially those newly graduated from college, do attempt to apply research-based strategies, like the learning styles, with the intent of actively engaging students in learning. Most of these attempts to engage students fall short. The teacher becomes frustrated because the students do not respond.

The traditional school culture obstructs teachers' reflection on their practice. Thus, there is a gap between knowing the research strategies and successfully putting them into practice. The gap seems to be getting wider. Perhaps with more "soft skill" training teachers will be able to engage more students, and thus have better classroom management.

HOW WAS THE *MLSI* PROFILE DEVELOPED?

The *MLSI* is based on psychologist Carl Jung's (1923) work on the human mind. Jung discovered the way humans take information in and how they value the importance of the information. These discoveries developed into his findings of different personality types. Katharine Cook Briggs and Isabel Meyers piggybacked on Jung's work and expanded Jung's personality types, turning them into a comprehensive model of human difference which they then developed into their Myers-Briggs Type Indicator (1992/1968).

The next generation of researchers used the insights of Jung, Myers, and Briggs to help teachers reach more students. Bernice McCarthy (1982), Carolyn Mamchur (1996), Edward Pajak (2003), Gayle Gregory (2005), and Harvey F. Silver, Richard Strong, and Matthew Perini (2007) helped to convert the findings of Jung and Myers-Briggs into more practical and classroom-friendly cognitive diverse learning styles. Silver and Strong developed learning and teaching styles. The *MLSI* profile was developed with a focus on how students prefer to learn math.

The *MLSI* helps teachers implement the research-based instructional strategies they learned in their teacher-training programs. The question often asked is, "How can I, as a teacher, put this research into practice so that I can have my students attain a positive attitude toward learning math?" To support the plight of the teacher, especially the secondary teacher, there are statistics that loom about when students become disengaged in mathematics.

Harvey Silver, in his introduction to *Styles and Strategies for Teaching High School Mathematics* (Edward Thomas, John Brunsting, and Pam Warrick, 2010), wrote about the research he conducted in the classroom with teachers and students. Silver found that by third or fourth grade almost 80 percent of students have positive attitudes toward mathematics and feel confident in their ability to succeed.

By freshman year in high school close to 50 percent of the students have developed an aversion to learning mathematics and don't believe that they are "good" in math. As a result, these students incur a goal to take the fewest number of math courses they can in high school and beyond. This leaves 50 percent of the student population believing that math is only for especially talented students and those who want to enter STEM careers (Science, Technology, Engineering, Mathematics), leading to the general belief that mathematics is not necessary in life.

Silver (2010) stated,

> This idea should give high school teachers of mathematics the shivers. We know that mathematics is the heart of so many things that affect everyone, from economics to technology, from the complexities of global marketing. . . . Mathematics opens up career paths, empowers consumers, and makes all kinds of data meaningful from basketball statistics to political polls to the latest trends in the stock market.
>
> Quite simply, we cannot afford to have so many secondary students walk into a fast-moving technological society looking to avoid confrontations with mathematics. For if we send an army of math-haters out

into today's competitive global culture, we are short changing millions of students by severely limiting their chances of future success.

Today, with the implementation of the Common Core State Standards, students' dissatisfaction with learning math might be higher than 50 percent, and may be experienced at an earlier grade level. Complicating the intent to improve the rigor of a math program is No Child Left Behind (NCLB) legislation (NCLB replaced by the Every Student Succeeds Act [ESSA], 2015, https://www.ed.gov/essa?src=rn). NCLB requires that all students experience the rigorous program. The "for all" requirement does not bode well with low-performing student populations. This creates an anathema for school districts with an academically compromised student population.

DELVING INTO THE *MLSI* PROFILE

If you are a parent whose child is at the beginning of the thirteen-year math education journey, you may want to be proactive in your child's math education and focus on helping your child to develop a healthy, realistic, positive attitude toward learning mathematics. You can benefit from knowing not only your own math learning profile as well as your child's but also from being able to help your child navigate the school's math program throughout the thirteen-year educational journey and on to college and a career.

As a parent, knowing your own preferences as a problem solver can tell you a lot about how you learn best. A good starting point is to find your own learning style. As a parent of:

1. an elementary age child, knowing about preferences for learning mathematics will help you to evaluate the math program, and perhaps avoid issues with teachers;
2. a middle school student, knowing your own preferred math learning style as well as your child's may help with the challenge of the new math curriculum, programs, and homework;
3. a high school student, knowing your child's preferred math learning styles may help with filling any gaps in his or her understanding of math, with finding the best support services, if needed, and with developing a positive attitude toward learning mathematics.

To begin, review the sample questions from the following *MLSI*. The five questions have been selected from the *MLSI*. Even if you are a parent of a preschool or elementary age child, it would behoove you to use the profile to help you better understand how you prefer to learn math. You will be able to identify aspects of learning styles as your child grows up through elementary school.

It will also be helpful for you to evaluate how your child's math program will engage him or her in learning mathematics. You will be able to trace your child's math journey and make sense about what works and what doesn't in your child's math education. You also may reflect on your own childhood learning experiences.

Tips on taking an inventory:

1. There are no right or wrong answers, or good and bad results.
2. The inventory is like a thermometer and resonates with how you prefer to learn math at the moment.
3. Inventory results should be used for reflection, not prediction.
4. All students rely on all four learning styles to help them learn mathematics.
5. Students tend to develop strengths so that one or two styles may be much easier for them to use than the others.

The goal of the *MSLI* is not to pigeon-hole students; the goal is to help teachers recognize different styles of thinking and learning and to develop instruction that engages all four styles.

The directions and sample items that follow are all taken directly from the Math Learning Style Inventory for Secondary Students (Silver, Thomas, and Perini, Thoughtful Education Press, LLC, 2008, pp. 5–6, 8, www.ThoughtfulClassroom.com).

DIRECTIONS FOR RESPONDING— WHAT KIND OF PROBLEM SOLVER ARE YOU?

The *Math Learning Style Inventory* is a learning tool that provides you and your teacher with information on which learning styles you prefer the most and which you like the least when it comes to math. This information will help you and your teachers make better decisions about learning and teaching.

Preferences

The *Math Learning Style Inventory* is not a test. There are no right or wrong answers. If you're having trouble with a word, phrase, or idea, please ask your teacher for help.

The *Math Learning Style Inventory* is made up of twenty-two numbered statements about your preferences as a math student, followed by four choices, lettered A, B, C, and D. All you have to do is rank the choices in the order in which you prefer them. Use the following point system to rank your choices:

- Give your first, or favorite, preference—5 points
- Give your second preference—3 points
- Give your third preference—1 point
- Give your fourth, or least favorite, preference—0 points

Remember to assign a different number of points (5, 3, 1, 0) to each of the four choices in each set. Do not make ties.

Question 22. I like math problems best when they:

A. *Ask me to use logic to solve a challenging problem and to explain my thinking. See problem A.*

PROBLEM A

You are given eight golf balls. Seven of the golf balls have the exact same weight, but one ball is slightly lighter (you cannot feel the difference). You have a balance scale, but can only make two weighings. How can you find the lighter ball in only two weighings? Explain.

B. *Challenge me to use math creatively. See problem B.*

PROBLEM B

In geometry, two lines are parallel if they are in the same plane and never intersect. Take the mathematical concept of parallel lines and apply it to nonmathematical situations or objects. For example, two people might be said to be parallel if they live in the same town but never come in contact with one another. Think of at least three more examples of things that are, figuratively speaking, parallel. Be sure to explain how each example can be considered parallel.

C. *Involve real-life situations and problems people commonly face. See problem C.*

PROBLEM C

As class social chairperson, you order 256 T-shirts for the class trip. After checking the number of shirts carefully and placing them into the storeroom, you tell the section leaders to each pick up 1/4 of the shirts to distribute to their class section.

Tanya arrives first and takes 1/4 of the shirts. Later, Matt arrives and takes 1/4. During lunch, Rich stops by and picks up 1/4. Finally, just before the final bell, Nicky takes 1/4 of the shirts.

The next morning, you are surprised when Matt, Rich, and Nicky tell you that they don't have enough shirts. You can't figure it out—you know you ordered 256 shirts. Then you discover some shirts are still in the storeroom. Matt, Rich, and Nicky tell you they all followed your instructions.

What happened? How many shirts are still in the storeroom and how many do you need to give to Matt, Rich, and Nicky?

D. *Ask me to find correct answers. See problem D.*

PROBLEM D

Use divisibility rules to determine if the first number is divisible by the second number.

1. 1,075; 5
2. 699; 3
3. 385; 6
4. 117; 3
5. 3,242; 3
6. 2,002; 6
7. 13,766; 3

The four distinct math learning styles were first presented in the introduction. These four math learning styles are described below.

"MASTERY" MATH STUDENTS

Want to . . . learn practical information and procedures;
Like math problems that . . . are like problems they have solved before and that use set procedures to produce a single solution;
Approach problem solving . . . in a step-by-step manner;
May experience difficulty when . . . math becomes too abstract or when faced with open-ended problems;
Learn best when . . . instruction is focused on modeling new skills, practicing, and feedback and coaching sessions.

"UNDERSTANDING" MATH STUDENTS

Want to . . . understand why the math they learn works;
Like math problems that . . . ask them to explain, prove, or take a position;
Approach problem solving . . . by looking for patterns and identifying hidden questions;
May experience difficulty when . . . there is a focus on the social environment of the classroom (for example, on collaboration and cooperative problem solving);
Learn best when . . . they are challenged to think and explain their thinking.

"INTERPERSONAL" MATH STUDENTS

Want to . . . learn math through dialog, collaboration, and cooperative learning;

Like math problems that . . . focus on real-world applications and on how math helps people;

Approach problem solving . . . as an open discussion among a community of problem solvers;

May experience difficulty when . . . instruction focuses on independent seat work or when what they are learning seems to lack real-world application;

Learn best when . . . their teacher pays attention to their successes and struggles in math.

"SELF-EXPRESSIVE" MATH STUDENTS

Want to . . . use their imagination to explore mathematical ideas;

Like math problems that . . . are nonroutine, project-like in nature, and that allow them to think "outside the box;"

Approach problem solving . . . by visualizing the problem, generating possible solutions, and exploring among the alternatives;

May experience difficulty when . . . math instruction is focused on drill and practice and rote problem solving;

Learn best when . . . they are invited to use their imagination and engage in creative problem solving;

Let's look at the possible responses to the four selections for question 22.

Question 22. I like math problems best when they:

A. *Ask me to use logic to solve a challenging problem and to explain my thinking. See problem A.*

A student who has a preferred "Understanding" learning style would likely rate this problem a 5. The problem requires the student to explain his or her answer.

B. *Challenge me to use math creatively. See problem B.*

A student who has a preferred "Self-Expressive" learning style would likely rate this problem a 5. This problem invites the student to use his or her imagination. The student is encouraged to create three more examples of things that are "parallel."

C. *Involve real-life situations and problems people commonly face. See problem C.*

The student with an "Interpersonal" learning style will likely be drawn to this problem because it requires the student to apply math to real-world problems and would likely rate the problem a 5. Problem C invites a discussion.

D. *Ask me to find correct answers. See problem D.*

A student with a "Mastery" learning style preference will be able to use step-by-step procedures that were learned in doing previous problems using the divisibility rules. The problem would likely be rated a 5 by the student.

How does a standards-based problem from the 1989 *MathLand* curriculum program engage all learning styles? Let's refer to the "Hot Spots" map homework project in the introduction and evaluate how this assignment addresses each of the four learning styles. The homework "grid" assignment referred to in the Introduction includes a task that engages each learning style. See if you can identify in the problem the "hook" for each of the four styles (Mastery, Understanding, Self-Expressive, Interpersonal).

Finding coordinates of where you live would appeal to the "Mastery" student. Identifying the "hot spots" would appeal to the creative, "Self-Expressive" student. The "Interpersonal" student would appreciate the application of coordinates to real-world use and would look forward to the discussion of the maps brought into class by other students. And the "Understanding" student would like the opportunity to compare maps to other coordinate systems, such as an energy grid used by the power company.

You can use the learning styles to evaluate how your child's math homework and projects may interest him or her in learning mathematics. However, be aware that some problems may represent a style that does not fit in your child's comfort zone. It is important to work through your child's discomfort, and have your child realize the discomfort and how important it is to use all the styles in order to be successful in mathematics. Students with a dominant "Understanding" style may not like to explain their solution to a math problem.

All students need to communicate how they solve problems. There have been many discussions between parents and teachers about having students explain how they solved a math problem instead of just giving the answer. Merely allowing students to come up with the answer is not helping the student to develop their communications skills. Being able to explain an answer in a way that others understand the solution prepares students to be college- and career-ready.

See if you can identify the particular math learning style indicated for the answers to the following four problems selected from the *MLSI*:

Question 19. *I would prefer to demonstrate what I know about a math concept by:*

a. Doing a creative project
b. Reflecting on my learning in a journal
c. Completing a worksheet or taking a quiz
d. Conducting further research and writing an essay

Answers: A. Self-Expressive; B. Interpersonal; C. Mastery; D. Understanding

Question 18. *I learn best when my math teacher:*

a. Asks thought-provoking questions and lets me think for myself
b. Uses interesting problems to teach new concepts
c. Encourages me and my classmates to share our ideas
d. Gives me immediate feedback on how I'm doing

Answers: A. Understanding; B. Self-Expressive, C. Interpersonal; D. Mastery

Question 15. *In math class, the most important thing is:*

a. Being able to think "outside the box"
b. Sharing my successes and struggles with my teacher so I see how to improve
c. Calculating and computing accurately

d. Learning how to think and reason for myself

Answers: A: Self-Expressive; B. Interpersonal; C. Mastery; D. Understanding

Question 12. The best kind of math assignments:

a. Encourage teamwork and involve the whole class
b. Let me practice what I already know
c. Ask me to use data to prove something
d. Have interesting "twists" that make them unique

Answers: A. Interpersonal; B. Mastery; C. Understanding; D. Self-Expressive

A scored profile for the *MLSI* may look like this:

Mastery 95; Understanding 15; Self-Expressive 30; Interpersonal 58

The total points should add up to 198 for the *MLSI*. Depending on the student, the dominant style will have the highest score. All styles will have a score from most to least preferred style. Each student is different and therefore has a different set of four scores.

There is a comfort level for each style score:

90–110: A very strong preference; almost total comfort when using this style
65–89: Comfortable when using this style
40–64: Moderately comfortable when using this style
20–39: Little comfort when using this style
0–19: A very weak preference; uncomfortable when using this style

This student's dominant style "Mastery" score, 95, is very strong. His or her "Understanding" score, 15, is weak. The "Self-Expressive" score, 30, indicates that he or she will have little comfort when using this style. The "Interpersonal" score, 58, indicates moderate comfort when using this style.

What does this all mean in relation to your interaction with your child when they come home with a homework assignment?

1. "Mastery" 95: If the homework asks your child to use the divisibility rules, like *MLSI* problem D, he or she should likely have no

problem answering the question if he or she has the correct math class notes on divisibility rules.

2. "Understanding" 15: If the homework asks your child to figure out how to find the lighter golf ball with two weighings, as in problem A, he or she may have great difficulty completing the task and may give up.
3. "Self-Expressive" 30: If the homework problem, like problem B, asks your child to be creative and give three more examples, he or she may struggle to find nonmathematical situations or objects.
4. "Interpersonal" 58: If the homework problem is like problem C, your child may find the problem engaging if the problem can be talked through.

How does this play out in a traditional classroom for the child with the above *MLSI* profile? Traditional math programs focus mostly on "Mastery" (problem D) and "Understanding" (problem A) style problems. So this student might engage with the step-by-step procedural problems (divisibility rule), but could have difficulty when asked to "explain" how to find the lighter weight golf ball in only two weighings.

As a traditional math program advances, the student may do OK with knowing how to "solve" an equation, but not be able to understand how to interpret a data set representing the equation. In addition, in the traditional program the student rarely gets the opportunity to discuss equations and the application.

A standards-based math program, when instituted in the correct way, addresses all the learning style problems (A, B, C, D) in a safe learning environment. For example, the teacher may set up balance scales in class or at a learning station where students can spend time physically going over the golf ball weighing options.

Where does the typical student body lie as far their learning styles being engaged in the classroom? The traditional approach to teaching mathematics puts heavy emphasis on the Mastery and Understanding styles. However, we know that there are students who prefer to learn using the Self-Expressive and Interpersonal styles. This may indicate that half of the students in our math classrooms are never engaged in learning math using their preferred style. One wonders: Are we leaving many students believing that they cannot learn math or are not good math students simply because our classrooms don't engage their preferences?

In the elementary math programs this is not the case. Elementary schools focus on the "whole" child, and have more opportunities to create flexible schedules to address all learning styles. Students work in groups and have ample time to discuss solutions to math problems. The math content is connected and applied to the real world. Math makes sense. The disconnect begins as students enter middle school.

Reflect on your own school days and see if your style resonates with how you viewed learning mathematics. In the brief introduction to math learning style profiles, which learning style do you believe is your dominant style? Which learning style do you see your child developing? Do not be concerned if you cannot answer the math questions, this book is a "what does it look like" guide not a "how to."

TWO
Connects and Disconnects of the School Mathematics Program

What Are the Hurdles Your Child Will Experience in Learning Math in School?

Disclaimer: The topics in this chapter may seem lofty—different. It is okay if you feel overwhelmed. Students and teachers are overwhelmed, and in the dark. Please read the chapter with this mantra in mind: "It is that which we believe in that guides our decision making. If we are not clear on our beliefs in general—we make fuzzy decisions." Developing a belief about what math is and what it does, and how it relates to our lives, is key to learning as well as to teaching math.

This chapter is about hurdles students will encounter that will impact how they engage in learning math. Students may hate math for all the wrong reasons. However, knowing how they prefer to learn math will help them jump the hurdles. The relationship that a student has with his or her teachers may or may not be a hurdle. Chapter 3 will address the teacher hurdle, the importance of the teacher-student relationship.

Hurdle #1: How to Become a Student of Mathematics—
Setting Learning Goals

Learning how to study math. (How to become a student of mathematics.) What does it mean to be a student? Being a student of mathematics is knowing what is understood and not understood about math and having the perseverance to find some time to work on the areas that need work (concepts and skills) independently. Homework in *Navigating MathLand* will be referred to as independent work. A student who is proficient in math is able to set individual goals for what he or she needs to work on independently; then, after a math lesson, they can identify the areas of weakness on which to work.

The ultimate characteristic of a good math student is the ability to persevere in finding a solution to a problem. Most of the time, teachers assign a general homework assignment for all students. The assignment is not designed to address the weaknesses of each individual student. However, the student is expected to do the entire assignment, even if he or she already knows how to do a problem. Basically, homework (independent work) is hit and miss. Some students can do the assignment correctly; it is easy—but there is no challenge. Some students struggle with the assignment. One size does not fit all.

The teacher should identify the purpose of the "independent work" assignment. Are the students to practice a skill or apply a concept? The idea of the assignment is to challenge students to find correct answers. Parents often step in to "help" with the homework and wind up, in frustration, doing the assignment for their child. It is strongly recommended by educators that parents do not do their child's homework. That does not mean ignore your child's requests for help. There are many ways to help your child without finding the solution to the problem.

Let's go back to the parent who was away with her child on Columbus Day weekend, 2013 (from the introduction). The child brought his fourth grade math homework for the weekend. It was a place value assignment "à la Common Core." The parent and the child had no idea how to do the assignment. The child was not clear on what to do with the worksheet that was due the Tuesday after returning from the weekend. Maximo, Vilma, and John had a stressful weekend.

In this case the parent has the right to ask their child where were the notes taken in class? The notes should have a model problem, similar to the problem that was sent home. The child should be able to read and interpret the notes taken in class—not just copied off the board.

Having no lesson notes and no model on how to do the assigned problems is an example of the disconnection between the class lesson and understanding the rationale for doing the independent work. Good math students should be able to summarize what the class was about, what they need to do for homework, and how to access help if they are "stuck." Engaged math students are motivated to find solutions.

Students must learn to take meaningful notes on the lesson and be able to comprehend what the lesson was about. That means taking notes they can understand. If warranted, a call to the teacher can be made asking the teacher to make sure that student notes can be used to successfully complete the homework. Oftentimes, students who are slow in writing/copying notes miss the teacher's explanation of how the notes are connected to the lesson. Therefore, the student doesn't get a chance to clarify what they wrote down.

Learning how to take meaningful class notes is seminal to the student being prepared to do independent work. Students can then set goals for what they need to study. Independent work is where students hone their ability to persevere and find the solution. But first, students must be clear on what the problem is asking.

At this point, if your child does not have notes, a request for a parent conference would be timely and beneficial. There may have been notes provided by the teacher, but the student did not take them down, or was unorganized with filing each day's written interpretation of the math lesson.

There are alternatives to help parents review the math skills and concepts being taught. Parents who have no notes to work from may want to have their child explore information offered on the Khan Academy (https://www.khanacademy.org/) website. These online lectures are developed specifically for K–12 math lessons by topic. There are also apps for iPads that can help students. A good math student knows how to articulate what they don't know when searching in cyberspace for support. A prudent teacher would suggest that parents, when stuck on how to approach a problem, refer to online sources such as the Khan Academy for help.

In school, math lessons need to be formatted to address practices that serve as criteria for being a good math student. There are eight "math practices" upon which Common Core instruction is based. All practices characterize what constitutes how a math student needs to study math. The following are examples of "I will" statements that were aligned to the eight mathematical practices that foster the development of Common Core curriculum:

1. I will work on problems and not give up;
2. I will think about using words and numbers;
3. I will be able to explain my thinking to others and listen when they explain to me;
4. I will build with objects, draw pictures, and write with sentences;
5. I will use _____ as a tool to help me solve problems;
6. I will do my work carefully and question to see if my answer makes sense;
7. I will use the rules of math to make my work easier;
8. I will look for patterns in my work.

(Adapted from the Standards of Mathematic Practice © 2014 Common Core State Standards Initiative Questar III—Excelsior College)

The origin of these eight math practices will be addressed in chapter 5. Engaging math curricula (lessons aligned with the standards) foster all eight practices. Depending on a student's dominant math learning style, some practices are more favored than others. For example, a student with a "Mastery" learning style will like practice #7 (using the rules), and may feel uncomfortable with practice #8 (finding patterns.) As students become proficient in math, they learn to embrace all eight math practices, and they become aware of those practices that are more difficult for them to use.

What follows are some interesting anecdotes from the education field. In 1970, Linda Solomon taught a self-contained special education class in Gorton High School, Yonkers, New York. Linda believed that her students should take Regents level courses, like the students in general education. She taught her students the algebra curriculum and they sat for the NYSED State Algebra Regents exam in June 1971, just like the general education students in the school. To everyone's surprise, the passing rates for her students were better than the general education students.

She was questioned, as many teachers who go above and beyond are, about how she did it. What was her secret? Yes, over four decades ago there were teachers who believed in "leaving" no child behind before it became a law. When the Gorton administration asked what her secret was for getting her students to pass the New York State Education Department (NYSED) Algebra Regents, she said that the first thing she taught her students was how to be a student, how to study. Linda taught her class how to be math students.

In 2008, Doreen Sheldon taught science to a self-contained class. Her tenth grade students had to take and pass the NYSED Living Environment Science Regents exam; Doreen's students were always able to rise to the occasion. When asked what her secret was to get her students to achieve on such a difficult exam she replied, "My educational philosophy was to first teach my class how to be students." Engaging instruction provides students with the know-how and skills for being a student regardless of content area—be it science, math, art, social studies, and so on.

One of the engagement strategies suggested to teachers to help them gain insight into what math means to their students is to ask students to write their "math autobiography"—asking students to not only describe their journey but also to share their hobbies, interests, and how best they prefer to learn math. Instead of reviewing fractions on the first day of school, get the students to tell you what they learned last year and how they have viewed math since they've been in school.

The goal is to connect the students' interests with the math they will learn this year. Once a child knows how to be a student of mathematics, he or she will begin to relate math to their own life (viz., hobbies, sports, business, art, music, science, social studies, English, and career path).

Hurdle #2: Relating Mathematics to One's Life—
What Does Math Mean to Me?

Take a moment to look back on your experience learning mathematics. Try to come up with a definition of what math means to you and your education, career, and daily life. Are you able to state what you believe to be the "nature of math"? What are your general feelings or perceptions about math? Try to produce a list of words to describe math. If you were able to identify a math learning style (chapter 1) that represented you as a student, use your style to help define how math relates to your life.

For example, if you considered yourself as having a "Self-Expressive" preference in learning math, you might like nonroutine problems like the *MathLand* problem from the introduction—design your community. Think back to your math classes and your math teachers. Did your preferred learning style help you "fit" into your math class? If you believe you have a "Mastery" or "Understanding" dominant math learning style, you probably were a good fit for a traditional school setting.

Was your preferred way to learn math addressed in the math instruction? If you knew then what you know now about how you prefer to learn math, could you have excelled in math if you were allowed to have more problem-solving discussions with your fellow students? If you lean toward the "Interpersonal" preferred learning style, you may not have been a good fit in class because you rarely had the opportunity to discuss math with others. You might have been chastised for "talking" in class. Could you have gone further with your math courses?

Based on the criteria for preference in learning math you learned about in chapter 1, can you develop a belief as to what math is and what it does? If you cannot articulate the relevance of math to your life, don't give up. Persevere. It takes time to make sense of a subject that vacillates from the concrete to the abstract.

No other subject is as abstract as math and at the same time depicts reality. The dual nature of math often leaves students confused. Keep in mind there is no one correct answer for the definition of mathematics because each individual interprets math differently—based on how he or she relates math to his or her own life. What math means to your child is a great discussion to have throughout his or her thirteen-year journey through school.

As we introduce math in school to K–12 students, the "nature of math" needs to be presented in a way that allows them to understand how math relates to their lives. That is where math is abstract, concrete, and applicable to solving "real-world problems"—whatever the problems look like. Math is hidden in our daily lives and we need to apply math to function in society, whether in sports, cooking, skateboarding, driving, buying a cell phone plan, or choosing a career.

The story that follows is an example of how the confusion about math influenced the choice of career. A technology teacher was a participant in a geodesic dome workshop presented at a STEM conference in Oswego, New York. The workshop explored the knowledge of math needed in order for a builder to construct a geodesic dome.

The teacher expressed a desire that the math be presented to him, as a student, from a practical, applicable point of view, like what was needed to build a dome. It wasn't until he reached college when he studied to be a technology teacher that he realized that he could do math. He discovered in college that he wasn't as bad in math as he had surmised from his secondary school learning experience. He reflected on his high school years, remorseful about how, in all that time in school, he didn't think he was capable of being a good math student.

By the time a student graduates from high school, they should have a sense of what math is and what math does. More often than not, this does not happen. Another hurdle for students is understanding how math relates to their daily lives and eventually determines their career choices. Students become disconnected from learning math because math does not make sense to them and is not applicable to their daily lives. Often a student chooses a career based on the level of math "not learned."

So, why study math? For 10 percent of the student population, those with an "Understanding" math preferred learning style, what math is and what math does makes sense. What about the remaining 90 percent of the students? What does math mean to the majority of students? Does math make sense? Is the Common Core "new" math? By placing math "reform" on a timeline, it will become clear that there is in nothing "new" about the Common Core.

Mathematics is vast and has many interpretations, but the bottom line is that how well a teacher understands their own beliefs about mathematics has an impact on how mathematics is taught to your child. How often do you meet a person whose chosen profession is "mathematician"? Has your child come home from school and expressed his or her interest in becoming a mathematician?

Children meet doctors, lawyers, teachers, firefighters, electricians, and so on. These are real professions. Career days at elementary, middle, and high schools feature the usual suspects—doctors, athletes, actors, newsmen and newswomen, and politicians. Children also aspire to be President of the United States.

It would be a rare occasion to see at a job/career fair a mathematician booth. What you'll see represented are careers related to mathematics—such as programmer, statistician, certified public accountant, and perhaps physicist, to name a few. Surveys have identified kindergarteners' first career choice as a super hero or princess; a choice as unreal as being a mathematician. That is not to say, of course, that there are no career mathematicians. Most mathematicians work in academia as professors or for software companies. Just look up jobs for mathematicians online (https://www.mathjobs.org).

There is one career path clear to students who want to study math—to become a math teacher. Today there is a dearth of math teachers and school districts are always searching for "qualified" math teachers. Thus, the gate is open to come into schools for just about anyone who gets a certification in math to teach math.

HOW DO THE EXPERTS DEFINE MATH?

Referencing the experts can be confusing but provides good fodder for conversation. Is trigonometry important for getting along in our culture? By the end of a student's thirteen-year stint in public school, it would be growth-promoting for a student to understand what mathematics will mean to him or her. What math will students need to further their education or to choose a career? Would it also be advantageous for students to understand how mathematics influenced the doctrines of statesmen and theologians, astronomers and musicians—all of whom have shaped the course of modern history?

The primary disconnect for students, teachers, and parents occurs when there is no differentiation between what mathematics "is" and what mathematics "does." Students should also have been exposed to viewing how mathematics serves physical and social scientists, philosophers, logicians, and artists with the intent of demonstrating mathematics to be both abstract and tangible. Richard Courant, who coauthored *What Is Mathematics?* in 1941, purports the clearest view of mathematics:

> It doesn't matter what mathematical things *are*: it's what they *do* that counts. Thus mathematics hovers between the real and not-real; its meaning does not reside in formal abstractions, but neither is it tangible. This may cause problems for philosophers who like tidy categories, but it is the great strength of mathematics—what I have elsewhere called "unreal reality." Mathematics links the abstract world of mental concepts to the real world of physical things without being located completely in either.

In and out of reality, questionable links that connect reasons to study math, and confusing math instruction impact students' ability to make sense of what math means to them.

> Formal mathematics is like spelling and grammar—a matter of correct application of local rules. Meaningful mathematics is like journalism—it tells an interesting story. Unlike journalism, the story has to be true. The best mathematics is like literature—it brings a story to life before

your eyes and involves you in it, intellectually and emotionally. (Stewart, preface to second edition)

In fact, New York State brings the literature "story" characteristics to the latest machination of the math curriculum. EngageNY divides the the NYSED Common Core Learning Standards for Mathematics into three stories: "A Story of Units" for grades K–5; "A Story of Ratios" for grades 6–8; and "A Story of Functions" for grades 9–12.

The lofty goal of our educational system is to provide our students with an understanding of how mathematics is an integral part of the modern world. It is one of the strongest forces, shaping thoughts and actions, and is a living body inseparably connected with, dependent upon and, in turn, valuable in all other branches of our culture.

The idea of teaching mathematics like literature to involve students intellectually and emotionally is found in the "Standards Movement" that has been around for over forty years—with or without the Common Core. Students sitting in math classes should not only be introduced to formal mathematics but also come away with an understanding of meaningful mathematics—meaningful to their lives.

As early as 1967, Morris Kline wrote in his book, *Mathematics for the Nonmathematician*, about the new ideas that make math a human creation.

> Perhaps we can see more easily why one should study mathematics if we take a moment to consider what mathematics is. Unfortunately, the answer cannot be given in a single sentence or a single chapter.... One can look at mathematics as a language, as a particular kind of logical structure, as a body of knowledge about number and space, as a series of methods for deriving conclusions, as the essence of our knowledge of the physical world, or merely as an amusing intellectual activity. Each of these features would in itself be difficult to describe accurately in a brief space.
>
> Because it is impossible to give a concise and readily understandable definition of mathematics, some writers have suggested, rather evasively, that mathematics is what mathematicians do. But mathematicians are human beings, and most of the things they do are uninteresting and some, embarrassing to relate. The only merit of this proposed definition is that it points up to the fact that mathematics is a human creation.
>
> A variation on the above definition which promises more help in understanding the nature, content, and values of mathematics is that mathematics is what *mathematics does*. If we examine mathematics from

the standpoint of what it is intended to and does accomplish, we shall undoubtedly gain a truer and clearer picture of the subject.

Kline brings up the "So why study math?" question because he believed that mathematics is concerned primarily with what can be accomplished by reasoning. And here we face a hurdle. "Why should one reason? It is not a natural activity for the human animal" (Kline 1967, p. 30).

Kline cites that humans do not need reasoning to know how to eat, or to discover what foods maintain life, how to feed, clothe, and house themselves, get along with the opposite sex, engage in many occupations, and even climb high in business and the industrial world. One's social position is hardly elevated by how much one knows about trigonometry.

Like Kline, "Why study math?" is a conundrum for most of the students in school. Most students go through the motions to learn math. They realize that they need math to graduate, it's the means to an end. Kline's opinion is that math is not a content area that is necessary for survival. Unfortunately, in the twenty-first century it is necessary to learn mathematics for academic survival, like learning to drive a car may be seen by students as a necessity to survive. Over seventy years, Currant's and Kline's opinions about mathematics differed as to whether or not math is necessary for survival.

Hurdle #3: Ability Grouping versus Cohort Grouping

There is a movement among parents to hold back their children from starting kindergarten. Why delay a child starting school? It has been shown statistically that the older a child is in his or her class, the more successful he or she may be in becoming President of the United States or becoming an ice hockey star player. The great educational opportunity in America is our compulsory education system that welcomes all students into school, even if they are immature and cannot sit in their seats. Mature and immature students are all included in the same cohort (for example, the graduating class of 2030).

Everyone starts out with a tabula rasa. As the NYC marathon runners start out in waves with common times, our students all start from the same line. However, there is a difference in the criteria for running and for starting school. The runner who expects to finish the marathon in five hours will register for the five-hour wave. Ability and readiness to run the marathon is the determining factor for the wave, not the age of the runner. The runner starts off with the five-hour group, but may run faster and catch up to the four-hour group or slow down and come across the finish line in six hours.

Not so in public school education, where the starting line, or wave, for school is based on age rather than abilities. Students who start school all have different abilities. The cohort (for example, kindergarteners who were born between January 2008 and September 2008) all began school in September 2013. This cohort model creates an ability disconnect and a hurdle for students whose abilities are not addressed.

Here we mourn the loss of the one-room schoolhouse of the 1700s, where students of all ages were schooled. The one-room schoolhouse was replaced with the industrial model two centuries ago. There are no cohorts determined in the one-room schoolhouse. More information will be provided about the history of the evolution of the one-room schoolhouse in chapter 4.

Elementary schools are the starting gates. Prior to formal schooling, students receive life lessons from parents and guardians. Elementary school is the great equalizer. All students come to the race at different levels until they hit kindergarten. Kudos to the kindergarten teachers who work with all the students, getting them to jump over that first hurdle—acclimatization to what it means to go to school.

Cohorts create an instructional environment that makes it difficult to address all the levels of abilities that beginning students face. Here is where issues within the content areas begin. A child who excels in math may not get what he or she needs to keep them interested. All students of elementary age are curious and creative, but as children get older, somehow our system quells creativity in a one-size-fits-all curriculum.

There are public schools that recognize how cohorts limit students with different abilities. There are K–3 schools where students who enter in the first wave can be grouped, depending on their ability, in a kindergarten or third grade math class. There is also the possibility of pairing a struggling student with an older, more advanced student. The way math concepts and skills are reinforced is through peer interaction. Yes, the "Interpersonal" and "Self-Expressive" preferred learning styles are fostered in a mixed grade school environment.

Ability grouping and engagement will be further addressed in chapters 5 and 6. What can be done to foster a child's learning math? By the way, children are very sensitive as to the group into which they are placed. They know whether or not a fellow student is bright. It is easy at a younger age to begin self-efficacy modeling, where children know what they can or can't do and what they need to do to advance. Cohorts are hurdles, as the abilities of each child are not enhanced by the one-size-fits-all curriculum. By the way, at your place of work are all the workers the same age?

Hurdle #4: Teaching to the Test

Along with the rollout of the Common Core came the more rigorous state tests tied to the evaluations of teachers. Here is a perfect storm—the roll out of a "new" math curriculum, a "national" math curriculum, that has not shown either an improvement in test scores or in getting students to be more successful in learning mathematics. The "old" attempts at state math tests showed poor performance by students. The Common Core has shown that more students are underperforming in mathematics.

To intensify the storm, the Common Core mathematics curriculum has a more rigorous expectation for students to reach a passing level. As a result, more students do not meet state standards as evidenced in low passing rates on the state assessments. Coupled with a curriculum that is more difficult to teach, states are evaluating teachers on their students' performances on the exams. It is not a good mix as teachers get stressed out about their evaluations based on student achievement. Teachers then transfer their stress to parents and students, which in turn affects the classroom climate.

The outcome is that teachers are turning their classrooms into test-prep centers where students are drilled on exam questions. Students are constantly taking "practice" exams. To boot, high performance anxiety of a teacher is passed on to the students. NTI (National Teaching Institute) states that such an approach can be counterproductive, making learning math a painful process and increasing the anxiety about the importance of passing the test. By fourth grade, Maximo was experiencing math as painful.

There are alternate meaningful ways to prepare students for a test. These include providing strategies, skills, and techniques to help students study and effectively handle the test-taking experience. There are stats that show how much time is spent preparing students for high stakes tests—15 percent of the year for general education students which amounts to 27 of 180 days of prep, or five weeks of class instruction.

For at-risk populations, the stat is 47 percent of instructional time spent on test prep. That is almost half of the school year. So the students who need good, solid classroom instruction in order to come up to grade level are experiencing test prep instead.

NPR (National Professional Resources, Inc.) lists the four keys to test performance (Test Preparation: A Teacher's Guide [2010], www.NPRinc.com):

1. Mastery of Content
2. Effective Study Skills and Habits
3. Useful Test-Taking Strategies
4. Techniques for Regulating Emotions

Number one on the list is the focus of the traditional math curriculum. Mastery of the content has a huge impact on how the students approach tests. The more effective and skilled teachers are in "teaching" mathematics, the more successful students will be at "learning" mathematics. However, most of the time students do not have the chance to "master" the content—that is, by scoring over 85 percent on assessments designed to measure content skills and concepts.

But teaching only content and not addressing the three other factors affecting student performance creates poor student achievement. The skilled teacher also can prepare students to be successful in taking tests. Success is based on how well students internalize the skills and concepts needed to solve nonroutine problems. Here's a story about math teachers rising to the challenge of the state math test.

In 2000, in the Washingtonville Central School District in Orange County, New York, the NYSED grade 8 math scores were abysmal. The middle school math department was a professional and proactive department. However, the teachers and administrators were bewildered with student achievement on the NYSED state math assessment. At that time each math teacher had thirty-five students in a class, which included special needs students who were mainstreamed. It was a time when inclusion programs were becoming the forerunners of NCLB (No Child Left Behind) legislation.

At a department meeting, one of the math teachers mentioned that her own child's school was using a standards-based program, *MathScape*. The teacher raved about how the program engaged her child in learning math. Effective teachers offer the best suggestions for curriculum. The middle school was interested in further investigating the adoption of the *MathScape* program and decided to do a pilot.

The president of Washingtonville's teachers' union was a member of the middle school math department and she was supportive of the new resource. The program was ordered for the middle school and the department took advantage of the free staff development that came with the purchase (there were seven units for each level). It was the first standards-based program that the teachers had encountered.

The teachers embraced the program but found, like all standards-based math programs, that it was very difficult to get through all seven units in the first year. They were really worried about how not getting through the units would impact student achievement on the state test (a typical worry of math teachers not "covering" the topics). Their concerns notwithstanding, the student scores on the NYSED grade 8 math assessments went up a significant amount. The superintendent was impressed. What happened?

A standards-based program fosters understanding and mastery of concepts and skills—that is, math processes as well as content are addressed. *MathScape* engaged the students. The program provided meaningful activities that provided both practice of math skills and exploration of math concepts. There were lessons that required students to discuss their solutions to problems with their classmates. All of the four math learning styles are addressed in standards-based programs. Traditional mathematics programs focus mainly on "Mastery" and "Understanding."

Teaching to the test is the worst approach to student learning. Going over and over problems that will be on the test will not engage students. Students who know how to do a problem are bored, and students who never learned the concepts to solve the problem are demoralized. Washingtonville teachers taught the standards-based program; and even if they did not finish the program, the students became better problem solvers. The teachers did not spend a major portion of instruction teaching to the test.

What happens in classes that are designed for students who scored below level 3? There are statistics relating to how much instructional time the math teacher spends teaching mathematics, and how much is spent on teaching "to the test" or "reviewing" for the state exams. For example, does the math teacher with a class of level 2 students, more likely than not, spend 53 percent of the school year delivering instruction in math and 47 percent prepping children for the state test?

Thus, students who need instructional support wind up being given practice worksheet after practice worksheet, resulting in little or no progress since the children are not engaged in learning mathematics. The rest of the students who scored level 3 or 4 experience 85 percent of the year having math instruction and 15 percent spent on prepping for the test.

The teacher teaches to the test and math becomes more of a bore, using a prep program "workbook," rather than "engaging" instruction. You may want to ask your school district to share with you the number of students that, as a result of instruction, have moved from a level 2 to a level 3 on the state exams. How many students are now math proficient and have reached mastery (above 85 percent)? Inquire about the passing rate of high school teachers' classes on the Common Core Algebra I exams.

The teacher of the "slow classes" (levels 1 and 2) may not have been adequately trained or mentored to develop lessons for "at-risk" students. Adequate training involves teachers learning how to use continual assessment (formative and summative) to identify the gaps that the students have and then to prescribe an approach to remediate the student. Data on how individual students score on the state math exams gives baseline insight into where there are student gaps in understanding math concepts and skills.

The subtleties of the separation of lower and higher scoring students for math instruction may not have been as apparent in an elementary school setting, due to flexibility of the day's schedule to have students spend more time on math. Also, in elementary school your child has one teacher for the day. That helps make sense of content and often connects content areas for the students.

As your child advances through the school years, however, content areas are delineated into different classrooms taught by different teachers. No longer does the child have the advantage of having one teacher connect concepts in mathematics to science, social studies, music, art, literacy, and so on. As students advance from grades 5 through 12, the number of different teachers per content area increases so that by the time students reach high school, they might have as many as nine different teachers each day.

In high school, the math department may have four teachers teaching algebra; and each teacher might deliver instruction in a different manner. Your child in secondary school may have math early in the day or late in the day, or juxtaposed with lunch or gym. All these variables (location, schedule, delivery of instruction) impact how your child will process math instruction and prep and review for the test. In general, teachers teach to the test and often miss the instruction needed to support student understanding of how to approach the solution of nonroutine problems.

Furthermore, it is not fair for the students who still have to sit for a standards-based assessment in the spring of each year. Where there is a strong traditional math program, teachers rely on "practice for the assessments" resources, such as i-Ready and Castle Learning, to review assessment problems. However, this type of review for the test is futile if the students are never taught with a standards-based process approach. Students who have been robbed of engaging math instruction now have to sit through a review of tedious sets of assessment problems.

In the old days, reviewing test problems worked because the traditional content-based math program was assessed by traditional content-based assessments. It was easy for the students to identify the "types" of problems that were categorized, and to remember the steps to arrive at a solution. A content-based approach worked through the exam but the students did not retain the concepts or skills after the exam was over.

Reviewing old math assessment questions ad nauseam is prevalent in grades 3–12. It is a strong belief among administrators and teachers that spending a disproportionate amount of time reviewing old questions works, and students achieve passing scores. Even when the state assessment scores are posted as dismal year after year, the belief is that students needed more time reviewing test questions. Contrary to the ideals of review, students cannot reach mastery on standards-based tests if they are not taught the process of solving nonroutine problems.

It is important that students do have some idea about the type of questions that are going to be on the assessments. But standards-based questions are directed toward process and application rather than rote step-by-step. If students learned how to find the area of a circle in grade 7, they are not exempt from being able to use the formula to find the area of a circle to answer a grade 8 question on the state assessment, even if finding the area of a circle is not part of the grade 8 math curriculum.

Hurdle #5: School Year (Summer Break: Ten Weeks—Concepts and Skills That Are Not Instilled Are Lost)

Time, Time, Time . . . Basically, the issue of connects and disconnects focuses on time management related to the school schedule. All that the students learn in one year is often lost over the summer vacation, but not if the instruction was designed to have students internalize the math concepts and skills. The school day, the school year, the summer, after school—all are hurdles for the learning of mathematics. Connected with the time issue is a hidden hurdle rarely discussed, the old Carnegie Unit of study, where students are required to have class five times a week, 180 days a year.

The Carnegie Unit (Student Hour, Credit Hour) is a time-based reference for measuring educational attainment used by American universities and colleges. The Carnegie Unit is also used to assess secondary school attainment using a Student Hour (derived from the Carnegie Unit, per its original definition). The Carnegie Unit is 120 hours of class, or contact time, with an instructor over the course of a year at the secondary (American high school) level (https://carnegiefoundation.org/resources/publications/carnegie-unit/).

At the secondary level the Student Hour classes usually meet for forty minutes a day for thirty-six weeks per year. American schools typically meet 180 days, or thirty-six academic weeks, a year. (Note each quarter of a school year is typically ten weeks producing a forty-week school year. Four weeks are set aside, during the year, as vacation days.) A semester (one-half of a full year, twenty weeks) earns 1/2 a Carnegie Unit. To graduate from high school, students usually need to complete three Carnegie Units of math (Algebra I, Geometry, Algebra II).

These units came about during the late nineteenth and early twentieth centuries through a series of three disjointed events all designed to standardize the collegiate educational experience. The Carnegie Unit was designed to normalize admission to postsecondary education that, prior to the nineteenth century, involved a comprehensive examination, either by public oral process or private written process. The application processes were highly subjective in nature and varied greatly among the postsecondary schools. These admission methods were slowly discredited due to their poor reliability and validity.

The adoption of the Carnegie Unit by American school systems is owed to Charles W. Eliot at Harvard University who devised both a contact-hour standard for secondary education and the original credit-hour collegiate postsecondary standard. In 1894, the National Education Association endorsed the standardization of secondary education.

By 1910, nearly all of the secondary institutions in the United States used the Carnegie Unit as a measure of secondary course work. This was a result of the Carnegie Foundation (established in 1906) providing retirement pensions for university professors. Qualifications for a "Carnegie" pension were contingent on the universities enforcing the 120-hour secondary standard. Today, these units continue as the basis for evaluating student entry into college and for determining student completion of coursework and degrees.

There are educators who are critical of these units due to the arbitrary use of time as the basis for measuring educational attainment. The criticism is that student learning varies greatly even among teachers who teach the same content, such as math teachers. In the twenty-first century, the platforms for delivering instruction, such as distance learning and telecommunication, have provided opportunities for students to learn at their own pace.

And so the American public school system is bound by a nineteenth-century format—one that has not been reevaluated in over 100 years. Today, the students in secondary school are limited by the Carnegie system and cannot test out of a subject or course. Even if they understand a mathematics concept or skill, they cannot move on to the next level—they have to wait for the rest of the class. During class, valuable instruction time is wasted for such students due to the restrictions of the Carnegie Unit.

For example, there are numerous cases of students who have passed the Algebra I Regents exam in New York. These students have, for some reason, been unable to attend the class; they have lost "instructional time," often called seat time. As a result, they have failed the Algebra course. No student is barred (due to attendance) from taking the June Regents exam. Many students pass the exam without attending the course. In order to accrue a credit, or Carnegie Unit, for algebra, in the next school year those students had to "sit" through the algebra course again.

On the other end of the spectrum, some students may be able to complete the algebra course in half the year, yet they don't have the opportunity to accelerate their learning. The century-old overarching formula for assigning credit presents a hurdle.

Reading between the lines, if we decide to design credit-bearing courses differently, the economic determination of the number of faculty needed (the FTE) for the following school year is affected. Student enrollment for math courses could vary greatly, thus ascertaining how many math teachers will be needed to teach each course becomes difficult. It's another perfect storm in education while the Carnegie Unit remains status quo—not challenged.

The Carnegie Unit determines the length of the school year down to the number of periods courses meet during the week and the length of each period. The nine-period day has been created to make sure all of the content areas are taught. If a student has a forte for English (reading and writing), but needs extra time to understand math concepts, he or she is out of luck because English and math courses meet for the same amount of time . . . five periods a week for thirty-six weeks, or 180 days a year—one size fits all.

The school day also plays havoc with working parents, who need to provide child supervision for two-and-a-half hours every day after school, thus limiting the time that students could do independent work. Beside the daily time issues and the one-size-fits-all 45-minute period, there is the ten-week gap in the summer. The mathematics concepts and skills that occupy a child's short-term memory are lost. Only the concepts and skills that get internalized may be transferred over the summer to the next year's study of mathematics.

The inefficient use of time in providing instruction for students and the constraints of the structure of the school schedule cause students to fly under the radar. "Gaps" in understanding that are not detected early as well as missed opportunities for student acceleration create learning hurdles for all math students.

Hurdle #6: Grading

Grading in schools is the 800-pound gorilla in the room. How grades are assigned to students is very random, and there is no coordination within the school for grades. A student can earn an A ("excellent—great to have in class") as a grade with little or no clue as to what they can or cannot do, and without identifying areas in which the student is weak or strong. Sometimes grades reflect a student's effort to complete homework and participation in class. As a result students attain As and Bs but then fail the end-of-the-year assessments (state exams).

It is important that students be aware of what they can and cannot do. When asking why your child is not succeeding in math, do not accept the vague answer that he or she is lacking basic skills. Be specific and ask what the specific skills are that your child is lacking. Can the student multiply, divide, add, and subtract fractions? If the student is in high school, ask if the teacher has provided a calculator and showed students how to use the calculator to do arithmetic calculations.

Grades of A–F will not explain your child's strengths and/or weaknesses. The teacher should be able to show concrete evidence of math problems that your child has difficulty solving and to provide analyses of how your child approaches problem solving so that you can understand where there are misunderstandings. Most of all, your child should have a clear understanding of what he or she needs to do to attain proper solutions to challenging (nonroutine) math problems.

Each student is different and has different weaknesses. In traditional programs, "one size fits all" is the norm for the math class, which makes remediation difficult. There is often not enough time to cull questions about a procedure or skill that students may have. Especially in a class of thirty students, forty-five minutes leaves many students with little feedback as to what they need to do in order to improve. Homework is usually recorded as part of the child's grade as done or not done.

It should be clear at Open School Night in the fall how a student's grade is determined. Students receive their grade and believe that it reflects what they can do. The object of grading should provide a clear idea for students as to what they can and can't do.

WEB-BASED PROGRAMS TO EVALUATE STUDENT PROGRESS

To better understand the need for a more clinical approach to evaluating a student's math prowess, consider the following analogy. Let's suppose you are sick and you go to the doctor for a diagnosis. The doctor gives you the same medicine that other patients have been given because all patients came in with runny noses. Students get homework; everyone has the same assignment (take a decongestant for the runny nose). But what caused the runny nose?

There are, oftentimes, disconnects between the student's grade and what mathematics the student actually knows and is able to do. The hurdle is to get a clear picture of your child's performance. Often in math classes, the teacher decides to give the students the math test/quiz on Friday—the rationale being that if the test were given over the weekend the student would forget the material taught that week. The student will have a quiz or test (usually every two weeks) for the ten-week period and the five exams and quizzes will be used to "grade" the student.

Occasionally, the entire class underperforms on the test/exam. In this case, the teacher reteaches the material, then retests; placing a brake on the momentum of the course. Rarely is there a prescribed plan for each student's improvement. Often, teachers use quizzes and tests as threats of failure rather than assessments for improvements. For the middle and high school math teacher, the fact that they have 125 to 150 students a day makes it next to impossible to diagnose and prescribe a plan for each student.

So, the student who gets a good grade for math in grades 3–12 for each marking period may, in fact, be in danger of not passing the high stakes end-of-year math assessment. The teacher of forty years ago had to rely on formative assessments (classwork, quizzes, class discussion, unit tests) during class to identify the weaknesses of the students. There were no online programs to help assess each student's gaps in math concepts and skills.

As the American public school system moved into the twenty-first century, technology stepped in with normed assessments such as STAR (Standardized Testing And Reporting, http://renaissance.com) and aimsweb (http://aimsweb.com). The aimsweb program provides for universal screening, progress monitoring, and a data management system to support student remediation; it uses brief, valid, and reliable measures of reading and math performance for grades K–12 which can be generalized to any curriculum.

The STAR program provides interim data so teachers can set goals, respond quickly to student needs, monitor progress, and maximize growth. These assessment programs are used to identify students' weaknesses and align the weaknesses with the math standards de jour.

Teachers who are well trained to use aimsweb, STAR, or any other assessment tool can be proactive in identifying each of their student's math weaknesses and strengths. Ideally, students take the first STAR/

aimsweb assessment in September. The teachers receive immediate feedback for each student and can identify which math skills and concepts students need to work on.

The assessments can be administered again in January and March of the school year, or whenever the schedule permits. The beauty of theses assessments is that teachers have an ongoing current baseline of data providing useful information regarding each student's improvement. There are reports that can be printed out and sent to parents. Parents will know what their child specifically needs to work on to improve understanding of math concepts and skills. Teachers receive reports on the progress of classes and individual students. Here is an example of a STAR report.

Parent Report for John Doe

Printed: Monday, September 12, 2011; 9:12:15 AM
School: Maple Elementary School
Test Date: September 8, 2011; 10:28 AM
Teacher: Mrs. S's class

Dear Parent or Guardian of John Doe:

John has taken a STAR math computer-adaptive math test. This report summarizes your child's scores on the test. As with any test, many factors can affect a student's scores. It is important to understand that these test scores provide only one picture of how your child is doing in school.

National Norm Scores

Grade Equivalent (GE): 3.3
Grade Equivalent scores range from 0.0 to 12.9+. A GE score shows how your child's test performance compares with that of other students nationally. Based on the national norms, John's math skills are at a level equal to that of a typical third grader after the third month of the school year.

Percentile Rank (PR): 16
The Percentile Rank score compares your child's test performance with that of other students nationally in the same grade. With a PR of 29, John's math skills are greater than 29 percent of students nationally in the same grade. This score is average. The PR range indicates that, if this student had taken the STAR math test numerous times, most of his scores would likely have fallen between 23 and 33.

The STAR math test scores will help John further develop his math skills through the selection of materials for math practice at school. At home, you can help John develop his math skills as well. At this stage, he needs to work with numbers in the thousands, and practice multiplying and dividing basic facts.

If you have any questions about your child's scores or these recommendations, please contact me (the student's math teacher) at your convenience.

The jargon "data to drive instruction" has permeated professional development for math teachers. Basically, teachers are to use the technology (for example, STAR or Standardized Testing And Reporting) to develop baselines for each student in the fall of the year. Using the data from the fall assessment, they craft an improvement plan for the student, then implement the plan during the school year, stopping at midyear to reassess the student's progress.

Again, data availability without using it to improve instruction creates another perfect storm. Starting the school year with 150 students, coupled with time taken out of instruction to have the students sit for the assessment, doesn't allow teachers the time to review each student's data and make plans for improvement. If a student with gaps is sitting in a math class that needs to "cover the curriculum" for the high stakes test, these gaps will not be addressed.

The student is set up for low achievement. However, teachers are not given the time to analyze the data. Schools may have programs like STAR, but often the data is not used due to inadequate training of the teachers and a lack of time provided for teachers to fully analyze the data and revise their daily lessons. The data produced from these programs allows teachers to glean concept and skill mastery, a baseline for needed instruction, an instructional plan, and student assessments.

Hurdle #7: A Lost Year of Math Due to Poor Teaching and Poor Transitions between Math Programs

One of the greatest connects and disconnects is the good year/bad year learning experience. As we get older, for adults, time seems to fly by—a year might only be 1/55 of our lives. But for a student starting out who is nine years old, the next year represents one-tenth of his or her life. When a student has a bad year in school, any disconnect for learning math could prove a disaster.

The shift from elementary to middle school is a huge change. Students come from being the oldest in school to being the youngest in the middle school. Finding lockers, following a schedule, and dealing with subjects that have been separated into their own content areas taught by different teachers (math is now cloistered in its own area) can all contribute to apprehension for students.

Whether a student is in elementary, middle, or high school, having a "bad year" in math will impact a student for the rest of their lives. Maximo liked his third grade teacher and brought home good math grades. However, Maximo achieved a level 2 (failing) on the grade 3 state math assessments. There was a hidden disconnect between the math being taught and the math being assessed. Vilma, Maximo's mom, did not receive the grade 3 state math score until the summer.

With no math support to make up for the gaps over the summer (the grades on the exams are not readily available in June) coupled with a grade that did not reflect the student's achievement in mathematics for grade 3, Maximo entered grade 4 at a disadvantage. To make matters worse (and create the perfect storm), Maximo's fourth grade teacher announced to the parents that she was having difficulty with understanding the grade 4 Common Core math curriculum. Again, the issue of properly training teachers unfolds.

Maximo's grade 3 teacher may not have had access to any prior math data, and he or she was not able to have any formative and summative assessments in order to identify the student's "gaps." Therefore, the teacher was not able to have a conversation with the parents to provide summer work to address any gaps. Now in grade 4, both Maximo and his parents are frustrated.

To complicate matters, districts are always searching for the "silver bullet" math program that will guarantee improved student achievement on the state math assessments. Changes occur more often in the transition from the K–5 math program to a different math program in grades 6–8. As a result, students may lose a year of good math instruction and thus begin to develop gaps.

For example, if a student has not mastered fractions, the student's grasp of proportional reasoning is compromised. Students who lack understanding of proportional reasoning do not understand algebra; and gaps increase from school year to school year, the largest gap being in the transition from high school to college.

Many students who are proud of taking AP calculus as seniors in high school have a great deal of difficulty with college calculus (especially when the first year has been waved). As freshmen in college, even though they passed AP calculus, these students repeat the first semester of college calculus.

Many students who take precalculus in high school wonder why they can't even qualify for college-level math when they enter college, and therefore need remediation. This raises the question, "Are students prepared in mathematics (grades K–12) for success in college?"

CONGRATS! YOU NAVIGATED THE CHAPTER—LET'S RECAP!

Chapter 2 provides parents a lot of information to process. In summary, as children develop an understanding of what it means to be a math student (that is, to study math), they will internalize what math means to them. Another obstacle is developing an understanding of what math is or is not. This is not an instant task, and it may often lead to many "ah ha" moments. This is important because there is a general confusion in grades K–12 as to what math is.

"What is math?" will be an ongoing question for a student's thirteen years of school. There is no right answer, and the answer must include a relation as to how math is connected to one's life.

"Why Study Math and the Lofty Goal of Math Education?" is a hurdle that finds students mired in math wars about what is important to study, who is left behind or not, what part of the curriculum is important, and what parts can be skipped. A student who has not developed a personal belief as to why he or she should study math gets caught in the holes of the "Swiss cheese" instruction of a curriculum interpreted by the teacher.

For the shift to occur, important factors need to be considered, including the student's relationship to the teacher and the quality of the mathematics instruction.

Learning to study math, based on its practical use and its abstractness, is a major hurdle to jump. How to engage students to persevere in problem solving is in itself a challenge. Why? Because the eight math practices identified are not evidenced in daily instruction. The perseverance practice (number one on the Common Core math practices list) is not included in the teacher's daily lesson plan. What does perseverance look like and how do we get a student to practice?

In chapter 1 the reader learned how the learning profiles foster engaging students in math. Even with the profiles scored and evaluated, there remain seven hurdles to overcome. No research has been done as to what actually develops one's preference in learning math. By the time your child gets to take the preference profile, ten years and six grade levels (K–5) will have passed. That is a lot of time (especially in grades 3–5) to influence a child's self-efficacy in learning how to study mathematics.

Parents who have had their own bad experiences in learning math, and believe they have limited ability, may accept the fact that their child does not do well in math. On the other hand, there is a stress that is added when a parent is "good" in math and expects his or her child to have inherited the same math ability. Confusion about the nature of math, beliefs on how math should be taught and learned, and the stress of state exams all provide that "perfect storm" that interferes with teachers' engagement of students in math.

An optimum environment to learn math incorporates a balance of practice, understanding algorithms, and applying math to the world; and this may differ throughout a child's school years. Even if the math is connected to other content areas (science, social studies, art, technology) as in elementary school, in the middle school years the content areas become disconnected and separate.

The gap of the summer is another hurdle. Teachers believe they have to reteach in September what students may, or may not, have remembered. They often are without the advantage of reviewing state test data for each student. Most of the time, data is not available from the state until July, when school is out. A student who tested below grade level in grade 5 does not have a plan for remediation to address the gaps for May, June, July, August, and September. Five months have gone by with no remediation plan.

Due to the structure of the school year and the belief that there is limited time to teach, decisions may be made to omit strategies that could help students. Math programs that are dissected (Swiss-cheesed) do not scaffold tasks needed for student understanding. Even with ongoing assessment programs (for example, STAR/aimsweb), online math lectures (such as Khan Academy), and homework programs that give students immediate feedback (like those found at Castle Learning) to support the Common Core, math education has not been brought into the twenty-first century.

When a teacher opts out of "teaching to the test" for a majority of the school year, parents and administrators often put pressure on the teacher for lots of review for the test; again, a balance issue complicated by "test anxiety" on the part of the students, teachers, and parents.

Standards-based math curricula that have been aligned with the Common Core State Standards for Mathematics launch the math curriculum in the right direction. If children are afforded an educational opportunity to follow a math program, like EngageNY, for all thirteen years of school, they will be able to define what mathematics means to them. Standards-based programs foster the perseverance that students will need to develop to solve problems that engage them in learning math.

A teacher's mathematical philosophy will influence how he or she interprets the curriculum. Students may walk away from the school year with lots of practice but no connection to understanding an algorithm. More information about deciding how to deliver a curriculum is presented in chapter 3—about the teacher and instruction. What the teacher believes to be good instruction answers the question, "Why?" This is not to place blame but just to note that ongoing teacher training is key to bringing math instruction into the twenty-first century.

THREE
Teacher

Who Is the Most Influential Person in Teaching Your Child Mathematics?

The purpose of this chapter is to create a better understanding of who a teacher is, what a teacher does, how a teacher develops instruction, and how the professional identity of a teacher impacts their instruction. Remember *Navigating MathLand*'s mantra: "No One Is to Blame." Yes, there are perfect storms in parent-teacher-student relationships, and there are some solutions to prevail against the "bad weather." The goal of the parent and the student is to recognize how classroom instruction addresses learning styles—in particular, the dominant style of each student.

You cannot change the teacher. The teacher needs to want to change. Going to the principal to complain without a list of valid reasons and requests often has a negative impact on the student-teacher relationship. Chapter 6, "Next Steps," will provide you, the parent, with a win-win approach to improve the learning situation for your child. Attacking a teacher is only going to backfire and make it difficult on your child. However, you cannot let a poor learning environment go unaddressed for a year because that is a year of your child's life.

Teachers cannot be all things to all students. However, teachers can realize how students differ from their own learning persona. Like their students, teachers have dominant learning preferences for mathematics. Teachers also have dominant teaching styles that may or may not dovetail with a student's learning style. Trained teachers have the ability to recognize the different learning styles and provide instruction to engage all students.

Most of all, teachers need to believe in the instruction, whether or not it is different from the traditional "Mastery/Understanding" methodologies they have been indoctrinated with as students themselves and by the culture of the school. What is needed is to provide instruction for all four preferred learning styles. The standards-based (for example, Common Core) approach to learning mathematics has required teachers to shift their instruction to accommodate all four preferred math learning styles. How the shifts are implemented will be addressed in chapter 5.

There are various ways to enter the math-teaching field. For example, one participant in a doctoral research study about preservice secondary mathematics teachers had earned a BA in philosophy. His eyes were opened when he went to look for a job as a philosopher. The only positions a college graduate with a degree in philosophy could find were professorships at the college level, positions that required a doctoral degree.

The participant evaluated his transcripts and realized that he had taken lots of math courses and could become a math teacher. As a result, he decided to get certified as a math teacher, which required student teaching. The participant was part of the research study that was focused on how the student teaching experience impacted the instructional decisions of new secondary math teachers.

All of the participants in the research study who were interviewed were asked to define mathematics. What did mathematics mean to them? Of all the participants in the doctoral study, one preservice teacher could articulate what mathematics meant to him. It was the participant with the philosophy degree. He could define math and also give an example of his definition.

While it is understandable that preservice teachers just starting out their careers may not have developed a definition of mathematics, it was an eye-opener that in all of the ten teacher-training programs in New York State, not one required teachers to define mathematics, to establish a philosophy of mathematics, or to articulate what math meant to them. You may want to ask your child's teacher what his or her definition of mathematics is.

Please take time now to reflect back to your favorite teacher(s). Choose a favorite from one content area (social studies, science, art, music, etc.) or level (elementary, middle, or high school) and one favorite math teacher. Take time in the space after this section to write down all of the attributes of the teachers that made them special.

Hold on to your list. Read the following true story about a teacher, then revise and compare your list to the teacher in the true story.

WHAT DOES THE MODEL TEACHER LOOK LIKE?

Let's begin.

In the summer of 1956, Ludwig Kasal trailered the new sailboat he built for his family up to Lake George, New York. Trained as a sea captain and boat builder, he knew that his family not only enjoyed sailing on Long Island Sound but would also embrace lake sailing. Like the 30-foot sloop sailboat *Moa* he designed to sail on the Sound, the *Chuck-Lin* (named after his son and daughter) was a 20-foot sailboat he designed and built specifically for lake sailing.

In the next decade, Mr. Kasal would complete his doctorate in education and become a superintendent of a BOCES (Boards of Cooperative Educational Services) in upstate New York. He had a passion for teaching and, with his boat-building experience, rather than pursue marine architecture, he chose to become an industrial arts teacher. He was a brilliant boat hull designer. Boat builders from around the world would call him for advice. But he wanted to teach students how to build. He spent most of his tenure (twenty-seven years) as the woods/metal shop teacher at Edison Vocational and Technical High School in Mt. Vernon, New York.

As a "shop" teacher, Mr. Kasal never had his students build birdhouses; he always had students learn the trade of building real houses, elaborate furniture, and, yes, boats. In 1954, three sailboats were drafted and built in his woodshop. Mr. Kasal and a former student (who also became an industrial arts teacher) built the first two.

One day, later in the school year, a guidance counselor came to the shop door to deposit a student who was scheduled to drop out of school. The student, a child of a prominent physician, was placed in Mr. Kasal's class for temporary holding until he reached the age of sixteen, which was the legal age to drop out.

This was not the first student to be dropped off. The guidance department knew that Mr. Kasal's classroom was a safe place and that a student would not get into any more trouble that would disrupt the school day.

Mr. Kasal was a principal's dream teacher. Once the student was dumped in the woodshop (for the entire day), there were no more discipline problems with the student.

The educational issue with this particular student was that he could not read. Dyslexia was common, even seventy years ago. Chronologically, the student was a tenth grader. Today, as was common in the 1950s, students with learning problems are socially advanced from grade level to grade level. Students with dyslexia are usually bright and determined, and their intelligence is not measured by traditional school verbal and mathematics achievements.

The two sailboats (in various stages of construction) were positioned in the large garage adjacent to the classroom. Shops back then had spaces big enough for students to build large pieces of furniture. The student took one look at the boats and said to Mr. Kasal, "I want to build a boat like that." Mr. Kasal's reply was, "I will help you, but you need to learn to read blueprints as well as directions."

The student struggled but was so motivated that, despite his disability, he learned to adapt and read blueprints. The student did not drop out. He instead graduated high school and became a successful businessman.

Mr. Kasal believed that all students could learn. He believed in each and every one of his students and respected them as human beings. Whether they came from poverty, broken homes, or homes where both parents were professionals, Mr. Kasal could determine what students needed to do to master the skills and concepts of the wood shop curriculum.

As a result of Mr. Kasal's belief system, his students graduated high school and went on to become Annapolis graduates, multimillionaire builders, successful businessmen, teachers, and some administrators. Oddly, no student became a guidance counselor. Back then, industrial arts was considered a "dumping ground" and Mr. Kasal would often get those students who just needed to attend school until they were sixteen and could legally drop out. Miraculously, those students never did drop out.

What were Mr. Kasal's instructional secrets? Number one was how he built relationships with students. He never blamed the students because they came from poverty, had "bad" parenting, or learning disabilities. Teaching was, and still is, all about relationships. Mr. Kasal "knew" his students and saw potential in each one no matter how gnarly their character.

Mr. Kasal often took his work home with him. Evenings would find him talking about his students who seemed to be doomed by their life's situation, written off by parents, family, and friends. It was not uncommon to have students visit his home. Mr. Kasal had a full wood and metal shop in the cellar of his home. There he was always busy honing his skills as a master builder. A continual learner himself, he was knowledgeable about the "new" fiberglass material that had entered the building market.

Mr. Kasal never missed a day of school. In order to spend the summer with his family, he taught night school and gave clarinet lessons after school. He was an accomplished clarinetist and had the distinction of building his own clarinet. Teachers in the 1950s, 1960s, and 1970s had to supplement their meager salaries with after-school jobs. Mr. Kasal's wife was a homemaker, making him the sole financial supporter of the family. However, where else can you have such a rewarding profession and get paid for it?

Mr. Kasal's woodshop was the safe haven in which to place these castoffs until they were old enough to quit school. These students, nonetheless, were expected to be productive and work in Mr. Kasal's class. He was interesting, funny, and strong. As a demonstration for the class he would break two-by-fours in half with his bare hands. He was always interested in who his students were, their interests, their goals, their hobbies, and their issues. And in return, Mr. Kasal's students respected and revered him.

He had talked to students in a stern but gentle way. The students knew that he accepted who they were and where they were in their learning. There was structure to his lessons. A chart of student progress was placed on the classroom wall and students knew from the step-by-step index cards what they needed to do to master a skill. Mr. Kasal's class was collaborative, he knew that students teaching students was one of the best instructional methods.

Mr. Kasal let the students explore on their own; even if it resulted in making mistakes in building projects (making wrong cuts in wood or using the wrong tools to hone a wood surface). The students explored and learned from their mistakes. If a student had a different way of performing a task, Mr. Kasal would say, "Go ahead and try it; see what happens. Let me know how it turns out."

To every story there is an ending. The ending to Mr. Kasal's story is based on what every teacher's end result should be. Answering the questions: "How do you know your students are learning?" and "What are the criteria that we use to judge learning?" The answers comprise the end game of how well students internalize and apply the skills and concepts that were taught. Along with mastery of the subject, we want students to have developed the self-efficacy attributes of independence and perseverance.

To continue the story, Mr. Kasal would store the 30-foot Long Island Sound sailboat for the winter at a local boat yard in Mamaroneck, New York. One year in late fall, empty paint cans, discarded by boat owners under a shed where all the masts were stored, ignited by spontaneous combustion. In the 1950s boats were wooden (fiberglass was not on the scene), and a roaring fire destroyed the boats that were tucked away for the winter storage.

Luckily, a thunderstorm arrived in the nick of time. The torrents of rain put out the fire that came right up to Mr. Kasal's sailboat. The sailboat was saved, but the mast was burned beyond recognition. This did not deter Mr. Kasal, however, as he had the mastery skills needed to rebuild the mast. During the winter months, Mr. Kasal drafted plans for a new mast and kept the plans in his shop in school. He always showcased his skills as a builder by providing a demonstration of how to build a project from draft stage to finish.

This year he was demonstrating how to build a mast. He expected to go into school during spring vacation week to build a new mast to replace the one that was lost in the fire. However, during the vacation week, Mr. Kasal was bedridden with chicken pox; and while lying in bed, he worried about getting the boat ready for the sailing season.

Just then the doorbell rang. Mr. Kasal's children went to greet the person at the door. What a surprise! It was Mr. Kasal's students asking where they should place the new mast they had built for the boat. They had taken Mr. Kasal's drafted plans and built the new mast—completed, rigged, and varnished to a "piano-like" finish. End of story.

How do you know your students have learned? They are able to persevere, study a problem, and create a product based on their own understanding of the project. Mr. Kasal's students had learned; and they performed the task of rebuilding the mast independently, without Mr. Kasal's help or guidance.

Why use Mr. Kasal as an example instead of a math teacher? Regardless of the content, there are attributes that all teachers need in developing their teaching practice. Many of these teaching attributes are developed on the job. Student teaching does not come close to helping new teachers acclimate (more about teacher training in chapter 4). Industrial arts was a content area that embraced science, technology, engineering, and mathematics (STEM) and also arts (STEAM). The instruction in the woodshop implemented all the strategies that engage students in learning:

1. Challenging content that hooks and motivates students;
2. Accepting students where they are and building an individual learning plan that helps students set their own goals;
3. Providing lessons that allow students to explore, practice, and discover concepts through guided questions; and to collaborate with other students;
4. Creating a safe environment where there is an opportunity to learn from mistakes.

Good teachers believe that their students can learn. Teachers are masters of their content area, are diagnosticians, and most of all, they develop a professional relationship with their students where the teacher understands the preferences for how students learn and, like a doctor, can prescribe learning tasks germane to each student's academic prowess.

If you are looking for attributes of good teachers, look at Mr. Kasal. He never gave up on a student and provided the safest environment for a child to learn. He allowed students to fall off that bike, and to retake the driver's test. The curriculum he developed was challenging, and he expected all students to reach mastery. Mr. Kasal was able to demonstrate what mastery in woodshop looked like—boats, houses, furniture, and so on. Most of all, students became independent learners and became literate in working with metal and wood.

Math is not only taught in math class but is taught and applied in all the other content areas, including physical education and health. Mr. Kasal had to teach his students how to measure, use tools effectively, make scale drawings, represent objects in three dimensions, apply algebraic formulas, use geometry to build projects, and apply trigonometry to curved surfaces. Take the list of attributes of your favorite teachers—

those classrooms where you believed you learned the best. Compare the list to Mr. Kasal's story as a teacher.

There are many forms of teaching that take place in a student's life from day one. Family has a profound impact on the first five years of a student's life. In fact, there is some psychobabble that claims a child develops a learning style by age seven. Today, students enter kindergarten at five years of age, having been influenced by learning experiences outside of the traditional school setting. All students enter a grade at a particular age, but their learning personas are vastly different. They enter a system that is not equipped to deal with the differences.

The first "teachers" children are exposed to are their parents, family, nannies, and friends. This nascent group is responsible for teaching children how to eat, go to the bathroom, exhibit appropriate social behaviors, to speak, and to trust or not trust others. And while some students have the luck of experiencing an excellent preschool training, most arrive with some deficiencies. All are expected to learn to problem solve, read and write, do math, and so on, but there are some basic needs that must be provided if students are to engage in learning.

Just like Abraham Maslow's pyramid of the Hierarchy of Needs (1943), before children can self-actualize morality, creativity, spontaneity, problem solving, lack of prejudice, and acceptance of facts, they must be able to experience the following steps:

1. Physiological—breathing, food, water, sleep, homeostasis, excretion;
2. Safety/security—of body, family, resources, morality, health, property;
3. Love/belonging—friendship, family;
4. Esteem—self-esteem, confidence, achievement, respect of others;
5. Self-actualization—realizing potential, what a person can be, accomplish. (www.simplypsychology.org/maslow.html)

To get to step number 5, self-actualization, students need to be comfortable with the first four steps. That is, before learning takes place students need to have a learning environment that provides the first four steps (from pre-K through grade 12).

Let's return briefly to review Mr. Kasal's teacher attributes. His classroom was a safe environment where students were taught how to properly operate machinery needed to build projects. There was a sense of community; there were rules and parameters that were set in order to encourage belonging and to build professional friendship. And while Mr. Kasal was a stickler for classroom decorum, he would remind but never embarrass students if their behavior got out of line.

Students had distinct goals, and knew what they needed to do to become master builders. It was clear from the chart on the wall how they had progressed. There was no race to "finish" a project because progress was based on mastery. Yes, there were core projects to complete, but there were also advanced projects (such as building furniture). Students were able to realize what building in woodshop meant to them and where it would fit into a possible career.

Keep in mind that the students who came to Edison Vocational Tech were a diverse group. Some came with a purpose to learn and some were placed because the school system did not know what else to do with them. In every school district, whatever administrator is in charge (including guidance counselors), industrial arts (today "loosely" called technology) has been and is a dumping ground for the unproductive student, the student who doesn't fit into the traditional program.

Even if you are not a believer in psychobabble, all teachers need to provide a safe environment in which students can make mistakes and learn from them. It is in this safe realm that students can self-actualize (that is, be creative, problem solve, and accept facts about mathematics). As a parent, you create an atmosphere of belonging to the family. The worst thing a student can feel is that they are not accepted by their teacher and/or fellow students.

Back to the teacher who is teaching your child how to learn math. Forty years of validated research shows how students learn math. It also shows how teachers' beliefs about how students will succeed in their math courses affects how those students will perform.

WHAT CHANGED AND/OR STAYED THE SAME — BELIEFS?

Until the introduction of the NCLB ruling, schools decided which students were capable of learning rigorous content. Students were homogeneously grouped based on verbal and math scores of standardized tests used to identify learning ability. It was easy for teachers to believe that all students could learn at their ability level.

But not all students took state exams like the NYSED Regents or SATs. Only the higher scoring portion of the student population was deemed worthy of taking such tests; and guidance departments played a key role in deciding which students made it and which didn't. Literally, a class system was determined by two scores, verbal and mathematical. If your child was fortunate enough to be in an elite position, then great. If your child did not "make the grade," there were always the blue collar options.

There were English language learners and those students from poverty-stricken areas who were identified by educators as being promising learners, but who were not being given the opportunity to take a higher-level curriculum. Thus, NCLB was proposed as a federal law as education entered the twenty-first century. Now, as Mr. Kasal had already done, teachers needed to change their beliefs and fully incorporate the tenet that all students can learn.

Teachers of the twenty-first century have to teach all students a rigorous math curriculum; this curriculum was always in place, but the prominent belief was that not every student should be expected to succeed. NCLB required all special needs students who are regularly assessed to take the same state assessments as the general education population. Those teachers, like Linda Solomon and Doreen Sheldon, who believed their students could learn a rigorous curriculum differentiated their instruction to meet the needs of their students.

Beliefs by teachers that their students can do the work and pass the test help them to design instruction that gets students engaged. There may have been teacher education texts written that have addressed the beliefs and understanding of how students prefer to learn mathematics. However, these "soft skills" texts, like preferences, are not part of teacher training programs. Content and instructional methodology may be the focus, but how to relate to students' preferences for learning mathematics has been placed on the back burner.

To illustrate the point, let's compare the teaching profession to the medical profession. When we go to a doctor or emergency room (even at the gym, yoga studio, or spa) we are asked for our medical history, what medications we are taking, what is bothering us, and how we are feeling. The medical profession takes us from where we are and what we perceive is physically wrong with us to the next stage of incorporating that information to diagnose and seek a cure for our ailment. Not so in math education. The curriculum is delivered without diagnosing our students.

Teachers rarely ask their students on the first day to list their interests, and write about their experiences with learning math. The "I hate math because" student response is an attitude that is rarely brought to a teacher's attention. From the start of school, the teacher has no idea about a student's past experience in learning mathematics. The data set (like a medical record) that accompanies a child as they transition through school often gets ignored, even though the information is readily available.

How a student has progressed in learning math is often lost as the student moves through the secondary portion of the thirteen-year journey. In fact, a secondary math teacher who is curious as to why a student harbors an attitude toward math would need to find the time to review the student's academic record. All teachers have the opportunity to go to their school's guidance department and review the student's folder.

The elementary record would be helpful because it provides a written report about a student's performance. A teacher's understanding of how his or her students learn provides clues to engaging them in the instruction. With all of the emphasis today on data-driven decision making, today's math teacher rarely has adequate time to look at how the student learns. The teacher is not to blame, as the data to be reviewed for each student is rarely available at the beginning of the school year. Remember, a parent can also request that a teacher review their child's academic folder.

"All children can learn" has become a cliché to some extent, but the concept is necessary in order to direct instruction. To pass judgment on a child's potential based on parent involvement and socioeconomic standing is misguided. Remember Doreen and Linda (teachers of special needs children); they believed that they could teach their students how to be students of math and science, which is an example of how a teacher's beliefs impact their instructional decisions.

BACKGROUND SUPPORT FOR HOW TEACHERS' BELIEFS IMPACT INSTRUCTIONAL DECISIONS

Preservice mathematics teachers that were participants in a research study of their transition into practice were asked how they came to the decision to study math. Most of the participants linked their interest to one math teacher from their middle or high school who had inspired them to study mathematics. Coupled with their interest in math, they were also asked why they chose to become math teachers.

Answers such as "I loved math and wanted to share that love with students" or "I love students," although altruistic, do not make a good rationale for becoming math teachers. The teaching practice is complicated and convoluted. Not one participant mentioned becoming an instructional specialist who could diagnose students' strengths and weaknesses in order to provide a prescription for academic improvement.

Not every child will love mathematics and go on to become a mathematician. The art of teaching math lies in getting students to not fear mathematics, and to realize that the higher the level of math courses they take, the more professional opportunities are afforded them, thus laying a foundation for advancement in learning mathematics.

It takes one teacher to inspire a student to learn mathematics. Does your child have access to that one mathematics teacher? Students want to know how mathematics applies to their lives. For example, there are mathematics teachers who believe that students need to learn how to mortgage a house. In reality, how many ninth through twelfth grade students will be purchasing a house in their high school years? However, if the students understand that knowing about a mortgage leads to a possible career choice, that might be the hook to learning math.

Students need to know how math is connected to other subjects, how knowing math can lead to a career, how math is applicable and can be fun; math can also be creative. Most of the time these applications of math are not clear to the students.

Parents often ask what they can do to help their child achieve in math, but this usually comes after the student is experiencing difficulty. Even if their child is experiencing success in math class this year that does not mean next year will be easy. Think again about learning to drive a car. People of all learning abilities know the importance of learning to drive a car. Wouldn't it be great to have students realize the importance of learning mathematics? As a parent, you provide the foundation for schooling.

UNDERSTANDING TEACHERS' PROFESSIONAL IDENTITY— BELIEFS AND PHILOSOPHIES OF MATHEMATICS IMPACT TEACHERS' INSTRUCTIONAL DECISIONS

As we go deeper into what math teaching involves, we will add to the list of teacher traits those that are specific to teaching math. For the sake of organization, the added attributes will be categorized as "professional identity." This professional identity is influenced by (1) teachers' beliefs about mathematics and how math should be taught; (2) how teachers reflect on their instructional practices; and (3) constraints that are inherent in the culture of the school district where they work. Take a deep breath—the next section is "heady."

Many philosophies about math exist. In relation to math education, there are three math philosophies identified by Dr. Paul Ernest that govern instructional decisions of math teachers: (1) Instrumentalist; (2) Platonist, and (3) Problem Solving. What do these three philosophies tout? How are they different?

An attempt will be made to present the philosophies so that they are clear to everyone. Think about any sport—baseball, golf, football, basketball, field hockey, and so on. Think about the "game," the rules for scoring, and practice. Sports analogies will be used to illustrate the three philosophies.

A teacher with an "Instrumentalist" philosophy believes that practicing unconnected skills is what math is about. Practicing skills is the way math is learned. It is called "drill and kill" in the education community. In the dark ages of math, math skills were practiced but rarely connected to reality. Think about the steps you need to do to find the answer to a long division problem (for example, 378 divided by 42). In the past, the steps to solve the division problems were assessed on a test. You had to know the steps (the algorithm) to get the correct answer. You did not have to know why.

But back then that was OK because the steps were assessed by tests—not the mathematics needed to understand the algorithm. Those students who excelled in the "skills" of adding, subtracting, multiplying, dividing, graphing, solving equations, and so on. were tested on those skills and received 100 percent. Voila! The students who passed the skills test perceived they were really good in math.

Consider taking an instrumentalist approach to sports—that is, by having the players practice the skills. In baseball, for example, students practice batting, then running sprints, and then fielding. If the coach only has the team practice and makes no connection as to how or why the skills are connected to the game, will the students understand the game?

A teacher who harbors an instrumentalist philosophy (mathematics is a set of unrelated but utilitarian rules, facts, and skills to be used for the pursuance of some external end) conducts practice lessons. If you think back to your primary years—and the three Rs—arithmetic was math. Not in today's twenty-first-century world, however. Yes, students need to practice their math skills but they need to understand why math works and then how to apply the math.

You may have started to learn how to count, and then to add, subtract, multiply, and divide whole numbers. The next level was a jump to fractions, decimals, percents, and some geometry. Did your teacher relate the fraction skills to the decimal skills? At best, each math practice was connected to some external end (finding how many buns could be distributed to 42 people if you had 378 buns), but the ends were not connected.

The next higher-level philosophy, or belief, in mathematics is the "Platonist" view. A teacher with the "Platonist" view sees mathematics as a static body of knowledge, a crystalline realm of interconnecting structures and truths bound together by the filaments of logic and meaning. Mathematics is not discovered but created. Too wordy a definition? Think back to baseball. Besides practice, a player needs to know the rules to play the game.

You might have experienced a teacher with a "Platonist" philosophy. Perhaps in algebra you created equations and graphed the equations on a coordinate plane. But there was still mystery shrouded by letters (variables) and algorithms that got in the way of making sense of what you were doing. You knew the steps on how to write an equation and graph it. The equation with variables was connected to a graph. Again, when you studied for the final exam in algebra you practiced many questions asking you to graph $y = 2x + 5$.

Your math teacher could make sense of linear equations, but you could not. But if you practiced the algorithms, did the sets of twenty-five to forty same-type problems every night, took the test made up of the same type of problems, you passed algebra—perhaps. But you were still in the dark (after doing all the procedures) about what linear equations were and why one would need to study them.

A teacher with a "Platonist" belief functions at the level of believing it is important to know and apply the "rules" of the game. The rules are the logic with which the sport is played. The rules unify and give meaning to the knowledge of the game, connecting the structure and truths of the game. In baseball, the rule is three strikes and you are out, no "ifs," "ands," or "buts." Three outs and your inning is over.

The rules help the player to connect practicing skills to the rules of the game. All that running around the bases that you did in practice helps you round the bases and make it to home plate before the ball reaches the catcher. A baseball player understands the rule that a ball batted out of the ballpark is a home run and they, hopefully, understand that they will have an easier time rounding the bases.

The third philosophy of mathematics is the "Problem Solving" view. A teacher who has a "Problem Solving" view considers mathematics to be a dynamic, continually expanding human creation, a cultural product, a process of inquiry and coming to know. It is not a finished product; its results often remain open to revisions.

You may have experienced problem solving in math when you were asked to apply skills and math concepts to real-world problems. Remember the two-train problem from your algebra course. One train leaves from the east, one from the west on the same track at the same time. One train moves faster than the other—figure out what time they will meet. The two-train problem was once considered nonroutine. There are several ways to solve this problem. After you had practiced solving the two-train problem, you were aware of the routine needed to attain the solution to two-train problems.

Mathematics, to the teacher who believes in the "Problem Solving" philosophy, is a game of strategies, used to solve problems. These strategies are open to revision. The "Problem Solving" philosophy underpins the "new math," the standards movement, and the Common Core. However, math reform includes practicing skills and knowing the rules for applying skills and concepts.

"Knowing the game," the strategies used to win baseball games, is akin to the "Problem Solving" philosophy. You often hear about talented athletes who can throw the ball 100 miles per hour. Yes, athletic wonders will win games. However, a player with expert skills may be a disappointment because there isn't a knowledge of the game. The pitcher has the skill, knows the rules, but does not know strategies, such as when (or to whom) to throw a curve ball, a fast ball, and so on.

Think back to your math education. You learned basic statistics—the mean and median as a measure of central tendency. You could calculate each value, and with a given a set of numbers, you could even graph the results. That is where the "Platonist" teacher stops. But what about learning the strategies for applying central tendencies (mean and median)? When is each central tendency used to best help you understand the data? The "Problem Solver" teacher leads students to use mean and median to best explain a set of data.

Recall those problems that showed patterns of numbers or shapes. They asked you to create a rule for the next term in the pattern that might be used. Learning strategies of the game are important in winning the game—just like learning strategies for solving problems are important to understanding mathematics.

Research shows that inquiry is the foundation of problem solving. However, our current school culture is cloaked in "Instrumentalist" and "Platonist" (maybe) views on mathematics. One of the monumental issues is to shift the beliefs of math education; otherwise, reform cannot happen. But shifting does not eradicate the need to practice and learn the rules of the game.

What about your child in this shift? Students who are natural problem solvers get the short end of math education. Especially those students who like to solve nonroutine problems. Often, the procedural approach to mathematics ("Instrumentalist") falsely gives students the perception that they are good math students. Recall from chapter 1 which math leaning style you believed was your dominant style.

If your dominant math learning style is "Mastery," you might feel more comfortable with the "Instrumentalist" philosophy. A dominant "Understanding" learning style would give you a proclivity toward the "Platonist/Problem Solving" philosophy. A "Self-Expressive" dominant style may feel comfortable with the "Problem Solving" philosophy, especially if students are asked to be creative in their solution.

Where does the "Interpersonal" style fit in? The answer is anywhere that the student can discuss and argue their math point and see realistic applications for mathematics. They will feel comfortable if math practice is connected to a real problem as well as the rule for a procedure; and they can strategize a solution with a group of students.

The hurdle for the students is that the majority of math teachers' beliefs are between "Instrumentalist" and "Platonist." A standards-based program is heavily focused on problem solving and having students learn different strategies to apply the concepts and skills they have learned. Many teachers do not believe in standards-based curricula, no matter how much staff development is done. A teacher's beliefs dictate the lesson, so students may only be given practice and rules but not strategies to problem solve.

TEACHING STYLES IMPACT INSTRUCTION

All math teachers have preferred learning styles and, like their students, have one dominant style. There are also teaching style profiles that indicate how teachers believe the content of their course should be taught. There are four domains, just like the learning style domains: Mastery, Interpersonal, Understanding, and Self-Expressive. As in the math learning style profiles, a teacher's teaching profile is represented by all four domains where one domain is dominant.

The teaching styles are reflected in the lessons the teacher develops. They are also reflected in the format in which the classroom instruction is delivered. Do any of the styles resonate with the math classes you experienced or with your favorite teacher? It is possible for a teacher to have a different teaching style profile than their math learning profile.

The Four Teaching Styles (copied from the *Teaching Style Inventory*, Silver, Hanson, and Strong, Thoughtful Education Press, LLC, 2005, p. 4, www.ThoughtfulClassroom.com) are depicted as follows:

Mastery Style Teachers focus on clear outcomes (skills learned, projects completed). They maintain highly structured, well-organized classroom environments. Work is purposeful, emphasizing the acquisition of skills and information. Plans are clear and concise. Discipline is firm but fair. Teachers serve as the primary information source and give detailed directions for student learning.

Interpersonal Style Teachers emphasize the personal and social aspects of learning, often by exploring students' personal life experiences and building feelings of positive self-worth. The teacher shares personal feelings and experiences with students and attempts to forge personal connections between real life and the content students are learning. The teacher believes that school should be fun and often introduces learning through activities that involve the students actively and physically or that allow them to work cooperatively. Plans often change to meet the mood of the class or the feelings of the teacher.

Understanding Style Teachers place primary importance on students' intellectual development. The teacher provides time and intellectual challenges to encourage students to develop skills in critical thinking, problem solving, logic, research techniques, and independent study. Curriculum planning emphasizes concepts and is frequently centered around a series of questions or themes. Assessment is often based on open-ended questions, debates, essays, or position papers.

Self-Expressive Style Teachers encourage students to explore their creative abilities. Insights and imagination are highly valued. Discussions revolvearound generating possibilities and finding new and interesting connections. The classroom environment is often full of creative clutter, while the curriculum focuses on creative thinking, moral development, values, and flexible, imaginative approaches to learning. Curiosity, unique and interesting approaches to problem solving, and artistic expression are always welcome.

In the traditional school setting, math is generally taught by teachers who employ a "Mastery" or "Understanding" style, but there are also teachers who employ both methods. Generally, the "Understanding" teaching style works well for honors students. The "Mastery" teaching style is the most traditional style and has worked up until now as standards-based curricula ask teachers to also include "Interpersonal" and "Self-Expression" style lessons.

The first group of students who lose out are the students who need some "Interpersonal" style lessons. The set of students who are "Interpersonal" learners are the students who get turned off to math. The "Interpersonal" learners also contain the highest number of at-risk students. The least addressed style, "Self-Expressive," leaves the creative student, the creative thinker, with no place to hone their skills. Those students are also turned off in a short time.

A master teacher is able to deliver instruction in all four styles. Check back to Mr. Kasal's class and identify how he incorporated all the teaching styles in his wood/metal shop program. The challenge is to get the math teachers to believe that all the teaching styles are valid and will work to engage students.

IMPACT OF THE SCHOOL CULTURE ON PLANNING ENGAGING MATH INSTRUCTION

There is not much that a teacher can do if they are trapped in a traditional teaching culture and have an open mind to standards-based teaching. There is also the issue of the department culture. If the department is traditional in its teaching practice, the teacher who wants to teach differently is often ostracized into being compliant with the traditional "Mastery" or "Understanding" style. There are some veteran teachers who might complain to the administration that "new" teachers are not complying with the traditional instructional styles.

This subtle "sabotage" is part of the political culture that thwarts any type of math reform. Shifting to a new math teaching style is one of the many hurdles encountered by schools. Again, the teachers have to believe that these new approaches work. Change is tough. More about the change is explained in chapter 4.

While a better understanding of the attributes of a good math teacher is important, there is little that a parent can do to change the system. Still, being aware that your child may have fallen between the cracks because his or her learning style is not addressed is half the battle. It is not that your child does not want to learn math, it is that he or she is not engaged in the learning.

For the preferred learning style that is addressed in a traditional curriculum, there is a danger that math students may appear to be learning when they are not. Real learning is impeded when there aren't enough nonroutine problems or projects where mathematics is applied. Remember that a doctor has a bag full of approaches to address medical problems. With the correct diagnosis and prescription, patients can improve. If no improvement is seen, the patient is then reassessed and a new plan created.

Math teachers who are aware of their professional identities can diagnose and prescribe what their students need in order to improve. However, many math teachers do not know their beliefs about mathematics, their math learning and teaching style preferences, or how to reflect on their own individual instructional practices. It is not their fault; it is just not part of the school culture or teacher training. The students (brilliant, creative, challenged) are the ones who end up with mediocre instruction.

FOUR
American Education System

The Nineteenth-Century Factory Model in the Twenty-First Century: Why Do School Systems Continue to Visit MathLand *but Never Stay?*

The purpose of this chapter is to help parents understand why math reform has not taken root. The result of failing math reform is the extremely slow implementation of twenty-first-century math programs, limiting instruction that could engage students in learning mathematics appropriate for the times. Readers will learn how the nineteenth-century industrial model of America's school systems hampers math reform and impacts students' thirteen-year educational journey. Addressed is how the development of math education for the twenty-first century is thwarted by stops and starts of "new" math curricula. The end result is a discontinuous curricular approach, K–12, to math instruction.

NOW

A Common Core math curriculum provides math instructional methodologies that provide the continuity needed for grades pre-K–12. The New York State Education Department math curriculum has been rewritten four times over the past thirteen years. The fourth rewrite is the current NYSED P–12 Common Core Learning Standards for Mathematics, adopted by New York State in January 2011, the standards document.

EngageNY is a body of math curricula documents created by New York State math educators as model lessons.

New York State decided to add standards to the pre-K and the elementary curricula. The additions to the Common Core standards are highlighted in yellow on the final document. Schools then had the option to use the standards document to write curricula that was designed to deliver instruction that addressed the clusters of standards in each domain. New York State Education Department developed suggested curricula in a module format for each grade level. EngageNY, the location for all NYSED Common Core resources, harbors every module that had been developed for math and English language arts.

New York State school districts have the option to adapt the EngageNY curricula to their current math programs. Some school districts have opted to implement the EngageNY math curricula as written while others have blended or modified the curricula. All schools that have adopted the Common Core should have their K–12 math programs aligned to their state standards document that is based on the Common Core State Standards for Mathematics.

Currently more than forty states have embraced the Common Core. However, across the nation, parents and teachers are working hard to overturn the 2011 standards-based initiative. States had the opportunity to adopt the Common Core, a national standards document, which provides a guideline as to which concepts and skills are to be taught in grades P–12 with the intent that if students change residence from one state to another state, they will be afforded uniform, standards-based instruction. The states also had the opportunity to revise the standards document by 15 percent to increase the rigor of the math programs.

As with other math program initiatives (that is, the "New Math" of the 1950s and 1960s, Back to Basics 1970s, NCTM Standards movement 1980s, "Fuzzy Math" 1990s), the Common Core may not get off the ground, but not for the reasons (too much stress on students, not developmentally appropriate for students) that feed the math wars. The archaic structure of the industrial model and the lack of teacher training create a perfect storm. As a result, the standards-based curriculum approach does not mesh well with the nineteenth-century industrial model.

Currently, parents have little control over the traditional industrial model for education (one-size-fits-all instruction) and the lack of proper teacher training. However, being aware of the systematic issues will help parents identify resources to support their child's study of mathematics while the system attempts to right itself like a listing ship. Again, no one is to blame. We need, as a nation, to release the nostalgic notion that school should still be like it was when our generation and our grandparents last attended school. Let's wake up! It's the twenty-first century and the world is different.

Students today need to compete in a global society. They need to be able to set goals, learn to self-educate, and be aware of business and career trends. Impacting education's entrance into the twenty-first century is the addition of the "cyber world" and social media. Each technological advance sets up a whole different playground for business and career choices. Today, schools tend to tune out the "cyber world" instead of embracing the power of all the online learning opportunities. But students coming into school have been brought up in a different learning environment so they become disinterested in traditional learning.

For more than sixty years, US educators have attempted to revamp how mathematics is taught. However, the math reform needed to connect learning to the demands of the global society has stalled over and over, leaving students in limbo and parents confused. Exciting and engaging standards-based math programs, like *MathLand*, *Investigations*, *enVision*, or *Singapore Math*, have been initiated in elementary schools only to be terminated in middle school because there is little or no training to bridge the transition.

It is important to understand the difficulties of implementing math reform for the American education system. In New York State, since 2000, the state education department (NYSED) has changed the math program four times. Prior to 2000, the high school math curriculum was algebra, geometry, and trigonometry. Then the new NYS school chancellor implemented Math A and Math B, which integrated all three subjects into two courses.

As a result of a poor rollout and ineffective training of teachers, not to mention a gender-biased Math A Regents exam, the state reverted to an Integrated Algebra, Geometry, and Algebra II Trigonometry curriculum (standards-based this time) in 2005. Six years later, New York State adopted the Common Core Learning Standards for Mathematics Pre-K–12 in January 2011, then developed the current EngageNY math curricula. Here we go again rolling out another "math curriculum." Case in point: For New York State in thirteen years, the math curriculum changed four times; thirteen years is the length of your child's educational journey.

All of these back and forth changes have been applied to a nineteenth-century agrarian school infrastructure. Understanding the education system du jour and how it impedes any reform is key. Why? Because during the thirteen years your child is in school, he or she will be caught up in the perfect storm of math reform.

Chapter 5 will address how the Common Core connects the math curriculum between the elementary and secondary schools. Idealistically, the Common Core is a rigorous math program needed to train students to forge the twenty-first-century global society; but logistically it has been a nightmare to roll out. So, while America is revamping its education systems, parents need strategies and suggestions on to how to fill the education gaps. Knowing the current system hurdles (that is, schedule, credit requirements) helps.

THE SYSTEM

The blame for poor results on international math tests (Third International Mathematics and Science Study or TIMSS) is often cast on parents, teachers, curricula, and difference in social and economic levels of students. It is important to understand that it is the structure of the American school system (the school year schedule, the daily schedule, teacher training, and how students are awarded credit) that has bogged down any progress in reforming math education. In America, we are aware that there is need for a change due to constant reminders by the media that students in Finland, Korea, and Poland are the top achievers on international math tests.

There is a catch-22. Despite protests against the Common Core, there exists agreement by all (parents, teachers, administrators) that we need to change the delivery of instruction. Educators know what needs to be done, but it's how it is being done that is suspect. The protests in learning communities are mostly attributable to how the Common Core is rolled out, and are rooted in a poor understanding about what the curriculum is all about.

Again, it seems that reform has done well on the elementary level, but it appears (from looking at the scores) that it is more difficult to transition the standards-based instructional strategies to the middle level. Therefore, the impact on math reform by the antiquated industrial model of our schools is least apparent on the elementary level but is most damaging as elementary students transition to the secondary level.

Little thought seems to be given as to how the infrastructure of the current industrial school-year model (ten months long, days that end at 2:30 PM, class periods of forty-five minutes, and the disconnection of the content areas) does not support standards-based curricula. The primary complaint by teachers is that there is not enough time allocated in the school schedule (only 45 minutes per day) to properly deliver instruction. This complaint will be addressed in chapter 5.

BACKGROUND HISTORY—WHAT HAPPENED TO THE AMERICAN DREAM? HOW DID WE GET TO THE FACTORY SYSTEM?

Let's begin with a story to illustrate what has happened to the American Dream. Charles Bosse's family came to the United States from Austria in 1890, when he was six months old. They lived in the German district (the upper eastside 80s) of Manhattan (New York City). Charles's mother was the superintendent of the building where they lived. Her job required her to, on a daily basis, clean the common bathrooms used by the tenants. Charles's father was a carpenter who built cabinets for the American Museum of Natural History.

German was the family language, but it was expected that Charles learn to speak English. German was spoken at home, but by the time Charles became an adult, he exhibited no trace of a German accent. Charles attended public school until the eighth grade, after which he took various jobs while he taught himself accounting. He was brilliant with numbers—a natural accountant. At the age of thirty-one he became the accountant for the vice president of the Union Pacific Railroad. That year he married his fiancée, eleven years his junior. As a gift for their wedding, the vice president gave Mr. and Mrs. Bosse a three-month rail pass to travel all over the United States.

At the end of the three-month trip, the Bosses had two weeks left, and the vice president suggested the couple take the train to Lake George, New York, deemed the most beautiful lake in the United States by Thomas Jefferson. By 1948, Mr. Bosse had become a prominent businessman. He could be recognized each day on the platform of the New Haven railroad (today's Metro-North) in a three-piece suit, complete with a gold pocket watch. That year he built a house on Lake George for his daughter, Gertrude Bosse Kasal.

Charles Bosse lived the American Dream. Immigrating to the United States, attending school only until grade 8, he was a "self-made" man who gave back to his community. He took the train from Pelham to work in New York City every day into his mid-seventies. Was Charles a product of his public school education or his will to self-educate to attain middle class status? Was America different back then?

Ludwig Kasal (the industrial arts teacher) was Mr. Bosse's son-in-law. Ludwig graduated from New Rochelle High School. Both Ludwig and Charles were the products of the American education system; both became successful professionals. However, in the eyes of his father-in-law, Ludwig was "only" a teacher. Teaching in 1948 was not a lucrative profession. Charles Bosse always wondered why Ludwig, with his great intelligence, settled for a teaching job. Ludwig Kasal was smarter; he did what he loved, and that was teaching. Isn't finding your career passion everyone's American Dream?

COMPULSORY EDUCATION—HOW DID THE CONCEPT EMERGE? HOW DOES "ONE-SIZE-FITS-ALL" EDUCATION AFFECT THE TWENTY-FIRST-CENTURY MATH STUDENT?

The American Dream has always been coupled with the American education aegis that all school-age children have access to a free education in public schools. Compulsory education in the nineteenth and twentieth centuries was the "perk" of being a United States citizen. A good education (academic and vocational) led to a well-paying job. America's idea to provide public schooling for the "common people" had seventeenth-century origins. The first model of a public school began in the 1630s, when Benjamin Syms left in his will the provision for the first "free" school to be created on his Virginia plantation.

Throughout the white population in colonial Virginia, children with wealthier parents received more schooling. Sons were better educated than daughters. Children that were orphaned, or came from poor families, had their future determined by courts, guardians, or masters. Educational resources were unevenly spread throughout the seventeenth-century white population of America.

In 1634, the orphaned and poor children in Elizabeth City County, Virginia, were given, via a private bequest from Benjamin Syms, property and livestock that was set aside for the first endowed free school in America. In England, free schools were established through philanthropy, and perhaps the idea traveled to colonial America. The Syms School and other free schools established in the seventeenth century were the origin of the American public school system.

By 1647 the Benjamin Syms School was in full operation mode. The school existed on 200 acres with a "fine" house and other accommodations, including forty milk cows. The term "free school" was interpreted to mean that the curriculum would be offered to all county children, regardless of economic status. A Common Core–type curriculum (the original three Rs) was included within articles of indenture for orphans and poor children to ensure that some measure of instructional continuity was required by law. Every child needed to learn to read and write.

And so the one-room schoolhouse was constructed; facilitating a concept that taught all ages of children, grades 1–6 (sometimes to grade 8) at one location. The dwelling measured twenty-eight feet long and sixteen feet wide, covered with heart-pine shingles. It had a crick chimney and two rooms that were lathed and plastered. The school was placed in an apple orchard that was fenced in to secure it against damages. The school operated from the 1640s to 1803.

TODAY'S INDUSTRIAL (FACTORY WAREHOUSE) MODEL SCHOOL

The industrial model of today took root in the nineteenth century. In 1843, Thomas Mann looked at the Prussian model of education, where students entered school in the same age groups. There were twenty-eight students in a class and teachers were trained. Thus, the industrial (or factory) model of our education system was launched. At the end of the twentieth century, with the invention of modern technology, it was thought that the industrial (factory) model of our school systems would morph into a better model suited for the twenty-first century. This did not happen. Schools are now operating on the same schedule followed in the late 1800s.

Today, compulsory education is more like compulsory attendance, with an infrastructure that has not changed. The results manifest themselves in extremes, such as students in high school who are made to sit through a course in which they have already passed the final exam but still need "seat time" in order to graduate. All students still need to achieve credits (Carnegie Units), as explained in chapter 2, which requires them to attend each course for five periods a week for forty weeks during the school year. Testing out of the course and getting credit for it was not an option.

Unfortunately, one mantra of American culture is "Our parents and grandparents came to this country and went through the school systems; so why can't our students do the same?" Why? Because the world has changed in 200 years—but not our schools! Ludwig and Charles could navigate success with an eighth grade education. An interesting point is that accounting and boat building were not part of the curriculum taught in either Ludwig's or Charles's public school. They had to find internships to enter their respective careers.

Like Charles Bosse, the Wright brothers did not receive high school diplomas, yet they engineered the first plane. Most modern inventions were fabricated outside of formal educational institutions. Styrofoam was invented in the corporate world. The inventors of the PC, Apple iPhone, and Microsoft did not finish their traditional college education, which leads one to question, "What does school have to do with careers?" According to Paul Simon, "When I think back on all the crap I learned in high school, it's a wonder I can think at all!" ("Kodachrome," *There Goes Rhymin' Simon*, 1973). School seems to get a bad rap.

In American education, the gap between our educational system curricula and learning the skills needed to become employable in the twenty-first century is widening. As a parent, you cannot expect that schools will fully educate your child in twenty-first-century skills—teamwork and collaboration, initiative and leadership, curiosity and imagination, innovation and creativity, critical thinking and problem solving, flexibility and adaptability, effective oral and written communication, accessing and analyzing information.

Teachers in our culture are demeaned, and teaching is still considered a job rather than a profession. Standards-based curricula are worthy attempts to bridge the gap but never get fully implemented. *MathLand* is visited by education reform, but the schools don't stay. There have been attempts to adopt the Common Core and to fully implement standards-based instruction in math classes. Schools try to transition to a standards-based approach, and to incorporate rigorous programs that are written with specific instructions to teachers as to how math can be taught to engage more students in learning.

Elementary schools have embraced such math programs but the standards-based initiative dissolves at the middle and high school levels. Where does the implementation falter? It falters in the system's inability to diagnose each student's math prowess. It fails by not providing rigorous training in the twenty-first-century skills needed to navigate a global society.

At this time in our history, with all the medical marvels and our ability to diagnose ailments, one-size-fits-all medicine is rarely practiced. Doctors depend on an individual's test results to prescribe a routine for improving one's health. It should be the same in education. With the technological advances today such as real-time diagnostics such as STAR, we can diagnose each student's strengths and weaknesses in learning, especially in learning math. One-size-fits-all compulsory education wastes time for students who already have demonstrated mastery and can apply their math course concepts and skills. It also becomes a waste of time for those students who have no clue about what or why they need to learn math concepts and skills. From both ends of the spectrum, the American Dream slips away. To the students in the middle, the curriculum does not challenge them to become better math students. With all the vacillations in math curricula and programs, who is impacted? . . . The students.

EMERGENCE OF THE MIDDLE SCHOOL PHILOSOPHY

The middle school philosophy came out research in the 1960s that supported the idea that students in a US middle schools were so swayed by their transition into adulthood that academics were secondary. It was thought that the middle school child needed to develop social skills and to get ready for the departmentalization of content in high school.

Middle school teaching teams were developed based on a child becoming socially appropriate as a young adult, in lieu of academics. "Kumbaya" moments were the focus. Teachers were selected based on how they could relate to the middle school age group (ages 11–14). Academic prowess of the teachers was secondary to providing a "feel good" culture that forgot academic rigor. When teacher candidates for middle school positions were interviewed the question was, "Would the newly hired teachers fit in and be able to function with their team?" Not, "Does the candidate know the math?"

Differing in structure from the junior high model, in middle school each student was assigned to a team of teachers that used a math curriculum possibly derived from the state's antiquated syllabus. Each team was assigned 150 students, giving each teacher an average class size of thirty students per five teaching periods (determined by the teacher's contract).

One difference was that middle school was typically comprised of grades 6–8, leading to reform wars regarding teacher certification (elementary generalist versus secondary content). Now the elementary-certified teacher, who was trained as an instructional specialist, was placed in a secondary setting with teachers who were content specialists. Also, there was parent outrage and gnashing of teeth because of eighth graders' potential negative influence on newbie sixth graders. Some districts went as far as including fifth grade in the middle school.

The bottom line is that there was a time in the 1970s when teachers in middle school were not responsible for helping their students pass state exams, leaving the high school teachers (who had to prepare their students for state exams in order for the students to receive a high school diploma and graduate) critical of the middle school education philosophy. Rightfully so as the middle school philosophy was not focused on academics but rather on transitioning students through pubescence.

It wasn't until the late 1980s (after *A Nation at Risk* was written in 1983) that the middle school educational culture was made accountable for assessing how students were learning (and understanding), not only mathematics but also reading and writing. But the stigma remained and the middle school "kumbaya" culture was profiled as being anathema to academic rigor. States began to develop assessments for math and English Language Arts (ELA) that were first administered in grades 4 and 8.

In the 1980s states began to address continuity in curricula at the district level with what were called "site-based curricula." Each district came to common agreement on curricula and assessments. The first shot at district-wide testing had growing issues. The local assessments were not normed and thus could not statistically validate that a math program was improving student learning. Slowly, states began to realize that the middle level math curriculum needed to be assessed, and that continuity in content, scope, and sequence needed to be developed to ensure that students were learning.

Decades later this gap in academic preparation, especially in students' learning of mathematics, tainted the integrity of middle school education. States started looking at the site-based middle school math curriculum as deficient along with the middle school cultural philosophy of selecting teachers based on their ability to relate to the students instead of their academic prowess.

To complicate matters, the middle school philosophy of not tracking fell by the wayside, and honors programs sprang up as early as grade 6. Starting in the 1990s, states began to administer math and ELA assessments to all grade levels 3–8, with the intent of identifying gaps in student learning. Inconsistency led twenty-first-century educators to develop standards-based Common Core math and ELA national curricula, with each state having the option to adopt.

Originally, honors level classes were not intended to be part of the middle school philosophical plan. But the "one-size-fits-all" traditional school program enticed parents to prevail upon the school district to create special classes. Remember, the traditional math programs were designed to focus on the middle student, leaving the bottom and top achievers in the class behind. Tracking is a nightmare for education and is still a staple of middle school math programs. Chapter 5 will explore how standards-based programs provide other ways to challenge the brighter student.

In middle school, students assigned to so-called lower level math classes will continue in the lower track for the rest of their high school years. They will never have the opportunity to collaborate with those students who have a better grasp on the math skills and concepts. Teachers who have the lower track often do not hold high expectations for the "slower" students. Students in the lower track generally receive classroom instruction 53 percent of the time with the remaining 47 percent being used prepping for the state test.

Students who have made it to the regular or honors level will have 85 percent of their instructional time dedicated to learning a one-year algebra curriculum. Those students will only have to sit through lessons relegated to review for 15 percent of their in-class time. The two-year algebra students, on the other hand, will experience 47 percent of the two-year program practicing for the algebra state test. Even after years of poor, almost dismal achievement on state assessments, the culture of school systems does not acknowledge that test scores are the result of a lack of math instruction.

This is how students are warehoused. Who gets stressed? The students and the teachers do. The lower track math students may not be able to take all of the courses needed to be college- or career-ready. Also, the data from the two state tests (grades 7 and 8) are rarely used to determine how to support students who all have different "gaps" in their math skills. Staffing decisions are made not based on student needs but on a formula. Once students are locked into a lower level math class, they rarely move into the normal school math mainstream.

THE STRUCTURE OF THE SCHOOL — DECIDING ON THE NUMBER OF TEACHERS NEEDED. HOW DO ADMINISTRATIVE DECISIONS AFFECT YOUR CHILD'S INSTRUCTION?

How does the administration decide on the number of teachers needed for middle school? It is important for parents to know that teachers at the secondary level are assigned classes of students using full-time employee (FTE) units — a formula used to determine the number of teachers. Usually, a middle or high school math teacher teaches five periods a day (out of nine periods). Therefore, if the maximum class size is thirty students, every 150 students require five FTEs (.20 per period) or one teacher (1.00). If your child is scheduled to take Algebra I in grade 9, he or she will be assigned an Algebra I teacher.

The number of algebra teachers will depend on the size of the school. With four or five elementary schools feeding into the middle or high school, a freshman class may reach the size of 600 students — requiring four FTEs' worth of algebra teachers which adds up to four teachers. Students are placed in classes as "bodies," not students. Any data regarding students' individual academic performances are not incorporated into the formula (note: grades and teacher recommendations are used to assign students to honors math courses, however).

When students get assigned, the individual needs of middle school students are not as paramount in determining their placement as they are in the elementary classes. Yes, the number of students entering elementary schools is considered for class size, but the elementary students' academic and social/emotional needs also play a major role in the teacher assignment equation. Instruction of students in K–5 is more tailored to student needs. Since there are more elementary schools than middle schools in districts, one teacher spends an instructional day with only twenty-five students in a class.

The elementary classroom model for K–3 has one teacher for twenty-five to thirty students who is responsible for all of the content areas (math, science, social studies, language arts). For the most part, the elementary schools are scheduled so that time on task, such as math, is somewhat flexible. Some elementary schools provide 60- to 90-minute math blocks of time per day. Elementary teachers pride themselves in understanding how children develop as students. Elementary teachers are great diagnosticians and can differentiate instruction so that students receive more individualized lessons.

Class size is valued more than student academic needs as far as assigning middle school students to math classes; one-size-fits-all for students who exhibit middle level ability and academic progress in mathematics. The honors students have already been selected using various criteria that are not consistent across the district. The placement of special needs students in math is determined by education laws based on IEPs (individual education plans). Individual student performance on state math assessments, rather than strengths and weaknesses, is not usually considered in student placement.

STRUCTURE OF THE CLASSROOMS AND SCHEDULE— NINETEENTH CENTURY

When you visit your child's middle or high school and classroom, does it remind you of your school? Do you think "Wow, there have been no changes in the way the classroom looks." Children still sit in desks (uncomfortable, hard desks of wood or metal) with small tops, and maybe a shelf underneath the chair to stack their books needed for the next class that will be in 45 minutes.

Elementary schools have cubbies for students that morph into lockers when they get to middle and high school. Perhaps you're saddened to see that there are no more black/green/white boards but something called a "SmartBoard" or "Interactive White Board." But you feel secure because the structure of the classroom has not changed—just like when you went to school thirty or more years ago!

On Parents' Night you might find yourself commenting to the teacher: "This classroom looks just like it did when I went to school!" Thirty years ago, the front of the room was where the lesson was launched and left-handed students didn't have the benefit of having a desk made for lefties, making it awkward for them to write. Now, think of entering a doctor's office or hospital still operating with the same equipment and devices from thirty years ago, with no disposable needles but reassurances from the nurse that the needles are sterilized—the same for tongue depressors—and cloth robes (not paper) that are washed and not disposed of.

With all the medical advances in the last thirty years, your doctor still believes in the old ways. Why? It worked in the war! In fact, your doctor has not attended meetings to learn the newest procedures. Why bother with what's new, take an aspirin and call your doctor in the morning. Would you say, "Wow, just like it used to be?"

Today in the twenty-first century, the structure of our schools, buildings, schedules, and classroom configurations all basically remain the same as they were in the nineteenth century. When new schools are built, they resemble the old school architecture and classroom design. Do you find it odd that there are no small rooms for conferencing or cubicles set up for students to do their independent learning? Even if there were a school designed for the twenty-first century, the plans would never get adopted because the state has to approve the building plan, and it has to be built to specific formulas to warehouse students.

The physical structure of the school does not allow students to find quiet places where they can reflect on the day, do independent work, or find out what they have done well in and what needs improvement. Each school day students, like lemmings, travel to at least seven different disconnected content areas, each area allocated 45 minutes to complete a lesson.

What if students need more time to grasp the lesson, or if a student already understands the lesson but needs the extra time to work on another subject? Are our daily lives cordoned off in equal periods of time? The ten-month school-year schedule has a two-month break in the summer and short days during the school year. Valuable time is lost.

COHORTS AND TRACKING—NINETEENTH-CENTURY CONCEPTS

All students still attend school based on their cohort (that is, their age in September of that year or whatever month is used to determine the cohort). In school, a student rarely works with a diverse age group of people as they will in real life, where people from different age groups work together. If a student happens to take an elective course (art, music, technology), he or she might be in a class of students who range from fifteen to eighteen years of age. Where do students get the opportunity to learn about working with a team that is diverse in age? How does one work with an older, experienced person?

Classes (cohorts) seem to take on their own personalities, deemed by teachers who look at the class "criminals, dullards, brainiacs" and judge the entire class by the few deviants or bright-lights. It seems inevitable that they do so because these "cohorts" travel together for thirteen years and each cohort leaves its mark on the team of teachers from their previous years who then "warn" the new team of teachers about the overall student quality of the group. There is no tabula rasa for the students who come though the thirteen years with reputations.

Compulsory math education was made worse by *Sputnik* (October 4, 1957) in that it started tracking. Up to that point, the natural tracking occurred as students dropped out of school to study a trade or enter the work force by grade 8 and later in grade 10. Students had options to get diplomas in high school—a local diploma or a diploma with distinction. In New York State the diploma of distinction was the "Regents Diploma." In the nineteenth and twentieth centuries, students could drop out of school and work, get a GED (General Educational Development certificate), or enter the armed forces to earn a high school diploma. Today, the world economy causes hardships for those students who drop out of school.

It was believed, in the 1960s, that only some students were worthy of learning all levels of mathematics. Those students who were deemed not able to understand algebra were placed in business math courses with the assumption that they would not go to college but instead would enter a business right out of high school. They were called basic math students. The second tier were the students who could tolerate some algebra, geometry, and trigonometry. These were students who probably could go on to a community college.

In New York State, the tracks of business and basic algebra (geometry and trig) were called non-Regents. There were more rigorous algebra, geometry, and trigonometry courses that students needed in order to take the SATs (formerly Scholastic Aptitude Test, then Scholastic Assessment Test, now officially just SAT). In New York these courses were called Regents courses that concluded with a Regents exam in June.

At one time, math teachers in New York State were identified as master teachers according to the number of students who not only passed the Regents exam but also attained high scores. The highest level in math was the honors course in which the elite students were selected to participate. These elite students were pegged as gifted by parents and teachers as early as third grade. By grade 5 the students were culled and directed to the best teachers and the best education. Through tracking, student populations were tiered into the elite 10 percent of the school and the other 90 percent. Thus the industrial model worked for at least 10 percent of the students.

All went well with tracking for about twenty-five years. In 1983, the US (federal) government released a report, *A Nation at Risk*, about the low scores on the math and reading national exams. The tome, coupled with the Carnegie Mellon *Science Matters* report, brought to light what students needed to know to be successful in the twenty-first century. In 1995, the United States was compared to other developed nations in the TIMSS report and placed, dismally, twenty-third among the superpowers.

The United States needed to prepare all students for the twenty-first century. Students needed to be literate in mathematics and ELA. With repeated dismal results on the TIMSS international exam, the Common Core curricula were developed. US governors agreed that the new curricula should be nationally uniform, but not all states ended up adopting the Common Core.

Today, in 2017, over three decades have passed and the United States is still "A Nation at Risk," ranked in the twenty-seventh to thirty-fourth position internationally when it comes to student competition in math problem solving in secondary schools. Compared to sixty-four countries, the United States is ranked in thirty-fourth place. Thus, it was a perfect storm that led to an "ah-ha moment" in the United States—that is, all students need to be taught rigorous courses of mathematics.

A noble thought, but the remnants of beliefs about what students can and cannot do based on tracking is haunting today's public school culture. The teachers and guidance counselors were aghast that students could no longer get math credit for "Consumer" or "Business Math." The National Collegiate Athletic Association (NCAA) now only accepts SAT math level courses on high school transcripts for athletes who apply for college sports scholarships.

The sports organization has clamped down and requires students to come to college having taken algebra, geometry, and trigonometry. What holds students back today is getting rigorous math programs established in schools where it is still maintained—by parents, teachers, administrators, guidance counselors, and others—that there are students who can't learn math. This belief usually permeates a district where the majority of the students come from low socioeconomic means and where math instruction engages, at most, 45 percent of the students.

How does this impact your child's educational journey? When a system does not believe students can achieve, their students will not achieve. Only 10 percent, at most, of the school population winds up in advanced courses. What happens to the other 90 percent? If students land in lower track math classes beginning in middle school, they will not have a chance to attain the math concepts and skills needed to meet twenty-first-century attendance requirements for college or for procuring careers.

TEACHER TRAINING

The introduction through chapter 4 provided an understanding about the way students can engage in learning mathematics based on their different math learning styles. You learned that teachers harbor the same math learning styles as their students. Teachers also exhibit teaching styles. Teachers' beliefs about mathematics and their individual dominant math learning and teaching styles all impact how meaningful math lessons are developed that "hook" students' interests.

A successful teacher is an instructional specialist who can evaluate students' levels of learning and prescribe instruction to ensure academic success. Addressing learning and teaching styles and developing the new Common Core mathematics standards-based curriculum are not enough to launch the math reform, however. All the training in the world will not assure that the Common Core curriculum will be executed successfully.

To better understand the system, let us begin with an analogy. In the last chapter we used the analogy of the medical practice to illustrate how a teacher is like a doctor who evaluates and prescribes instruction. In this chapter we will compare the American education system to the aeronautics industry, the initiatives of which focus on designing new planes and engines, as well as training pilots to fly the plane.

Using the aeronautics industry analogy, what US educators are doing is attempting to attach a jet engine to a biplane. Let's consider the biplane, invented by the Wright brothers at the turn of the twentieth century, as representing our current industrial model school system. Think back to 1903, the first successful flight at Kitty Hawk. Suppose the developer of a jet engine (representing standards-based math reform) came to the Wright brothers and suggested that the new jet engine be attached to the biplane. The engineers who developed the jet engine are convinced that the jet engine will assure that the plane will fly faster and at a higher altitude.

The promoters of the jet engine have developed a pilot's training manual that is 1,600 pages long and agree to spend five days in 1905 training the pilots to fly the plane with the new jet engine. The only problem is that the pilots are not convinced that the biplane will be able to fly with the new engine. There were no test flights of the jet engine that provided data to support this new plane initiative, and the pilots were not able to test the new "jet" plane to see how well it flew.

Today, the NYSED P–12 Common Core Learning Standards for Mathematics EngageNY curricula provide teachers with 1,600 pages, per grade level, of teacher manuals. Teachers are expected to read the manual and adjust their lessons to deliver the new standards-based math program with neither a test flight nor statistics to make the teachers believe that the shift will produce higher achievement on standardized tests. The teachers are skeptical and not eager to embrace the shift from the traditional teaching of math "content" to the standards-based "process" methodologies.

Complicating the training is the rollout of the new Common Core curriculum. The existing school schedule of 45-minute periods does not allow enough time for the standards-based programs to be properly implemented. Elementary school has a flexible schedule that is more suited for the Common Core. The teachers, the learning terminals for our students, are frustrated by the process and remain nonbelievers that the new program will improve learning in mathematics.

There is little time for the teachers to discuss (as a team) the new curriculum and teaching strategies. In essence, American educators are attempting to place a jet engine (the Common Core) on a biplane (the nineteenth-century industrial model). Chapter 5 will delve more into how the structure of the Common Core lessons makes it impossible to deliver in 45 minutes. However, the American education system has always played a pivotal part in actualizing the American Dream.

There are also disconnects in how math teachers are prepared in academia for implementing their practice. College students believe that they want to become teachers because they "love" children, "love" math, and look forward to the perks of teaching. They see numerous benefits: a job forever, medical, dental, pensions, opportunities to make six figures for ten months of work, and days that start at 8 AM and end at 3 PM. The newly graduated teachers have been indoctrinated in the factory model. They have spent almost twenty years of their lives attending nineteenth-century industrial model schools.

Teaching as a job certainly fits well with the factory model. Teaching as a profession, however, is lost in transition. Teachers as professionals are specialists who are able to evaluate students' educational needs and use various instructional strategies to convey the content of the course. Teachers, like doctors, need to evaluate students' levels of learning, identify their strengths and weaknesses, and prescribe what is needed to facilitate understanding. Teachers are instructional specialists. Training to become a teacher needs to be more like the training of medical professionals.

Inadequately trained teachers impact your child directly, as characterized by weak classroom instruction and a lack of student engagement. Student academic potential is not recognized and there is little feedback about a student's performance. On the positive side, school districts can opt to hire consultants to support teachers, also referred to as coaches. The best model for training teachers in practice is in the classroom on the job—simply, ongoing training.

In addition, as mentioned earlier, the school schedule does not provide the proper amount of time to effectively train teachers. Another example of a perfect storm is the practice by districts of scheduling one to three superintendents' conference professional development days in the school calendar—usually the first day after Labor Day, a day in late fall (Election Day), and then a final day in March. Teachers dutifully attend these workshops but oftentimes administrators do not provide the follow-up between professional development days needed to sustain the training.

Superintendent's conference days are usually a fun day for the teachers—"Entertain me, oh administrators, because I am going back in the classroom and doing what I want." Most of the responses from teachers on workshop evaluations are, "I would be better off marking papers today." Some teachers actually bring papers to presentations to mark while the consultant is presenting the workshop. The best time for teacher training is in the summer when teachers have no papers to grade. However, broaching this option to teachers, even with pay, does not fit well with the ten-month "job" expectation.

As just stated, since there is rarely follow-up to the professional days' program, what was presented to teachers on those days (folders and resources, etc.) is placed on desks, filed, or "circular filed." So if the consultant returns in November and asks the teachers to bring the materials from the first session, most of the materials are lost or forgotten.

DONNA DAVIS'S POINT OF VIEW ON BRINGING SCHOOLS INTO THE TWENTY-FIRST CENTURY—TEACHERS TRAINING TEACHERS

Donna Davis was hired to be the consultant/staff developer for Glencoe textbooks by the Harrison Central School District, Harrison, New York. The Louis M. Klein Middle School math department adopted a standards-based program in response to the edict of the superintendent who wanted the middle school mathematics program to dovetail with the elementary math standards-based program, a good idea that would provide a consistent pedagogical model as to how math should be taught K–8.

Adopting a standards-based program, getting teachers to buy into the program, and then getting teachers trained to implement the program are three different and difficult tasks for an administrator. The middle school math teachers were given the opportunity to select a math program, but it had to be standards-based. The department selected *MathScape* and Donna was hired as the staff developer. She spent thirteen days a year for two years working with the teachers to train them in how to use the program.

Donna has spent over ten years (1995–2012) working on staff development training with math teachers in all fifty states. Her focus was on teaching them how to use Glencoe's standards-based secondary math programs; for the middle school it was *MathScape*. Donna also presented workshops at state and national conferences on how to engage students in mathematics. Her retirement from Glencoe was her second time retiring. Prior to her employment with Glencoe, Donna taught secondary mathematics in the Baltimore city schools and was selected to mentor and coach math teachers throughout the Baltimore school district.

Donna was a top-of-the-line math staff developer with experience working with math teachers from New York City to Fairbanks, Alaska. She has seen and done it all when it comes to providing top-notch staff development. Not only did she work with teachers who believed in the program, she also witnessed negativity in teachers that caused her alarm.

With her talent and expertise, she has seen firsthand how difficult it can be to work with teachers to get a standards-based program implemented as designed. Donna's perfect storm hindering progress was that the schools hadn't come into the twenty-first century in the use of technology. She saw archaic school systems that resisted embracing the technology of the twenty-first century. A twenty-first-century school uses technology to provide instant feedback for students and to create individualized learning plans.

Textbook companies of today have embraced the use of technology. Schools have selected comprehensive textbook programs that include online resources and manipulatives, all aligned with math standards. In her training, Donna worked with teachers to show them to how to use technology resources with lessons that are equipped for the SmartBoard. Students no longer needed to haul books home, they can access their textbooks online. In addition, calculators have become very sophisticated and offer great support to secondary math programs.

Math teachers in general do not embrace technology or manipulatives. They explain that they try but the lesson is not immediately successful, so they revert to the old way—drill and kill. In teacher math workshops, there exists an undercurrent that questions whether the Common Core will last or if the states will renege on the initiative. In the summer of 2014, hundreds of teachers stormed the NYSED in Albany to protest the companies that are creating and piloting the state assessments; the bottom line is that teachers want to do away with standardized tests.

Just think if pharmaceutical companies wanted to do away with standardized tests to test products. What would happen with the quality of our drugs? Standardized testing gives us normed testing which in turn helps identify what students know or don't know. In chapter 3 the reader learned that data from the tests is very informative and can be used to drive instructional decisions. Testing, if done with the correct intent—to diagnose what students know and don't know about math—can synergize the learning process.

There are other "red herrings" that are waved in front of consultants by the education community regarding technology. Excuses range from low socioeconomic home situations to low abilities of the students entering schools. Poor students do not have computers at home, not to mention the risk of lost passwords to online resources. This leads to a general pushback to implementing student evaluation systems (systems that would provide ongoing data about students' strengths and weaknesses in math concepts and skills).

For over a decade, Donna Davis worked in 1,000 districts in every state in the country. Often, after the training on *MathScape* for a district, Donna would be asked by the administration to go in and see how the teachers she had taught were doing. The administrators asked Donna to observe the teachers and report back.

To her surprise, the teachers were still teaching traditionally—that is, rolling out the standards-based lessons like traditional lessons by implementing "Do Now" techniques first, then reviewing the homework, or demonstrating a sample problem using an algorithm for the students to practice and then assigning homework of the same type of problems. What was left out were the math practices of perseverance, argumentation, and the use of tools or manipulatives. You will learn more about traditional versus standard-based lessons in chapter 5.

Donna Davis has experienced this provincial attitude throughout the country. She observed that advances made in twenty-first century technology had not been embraced by school communities. She also agrees that schools designed for an agrarian society are a major hindrance to the progress of standards-based math programs, the reason being that the 45-minute period schedule does not allow students time to explore and discuss the math. This leads to the next problem—double time (the 80–90 minute "block" period).

TEACHING IN THE BLOCK— TRYING TO IMPLEMENT CHANGE IN THE SCHEDULE

Double periods (more time on math tasks) are one of the "out of the box" improvements in schedules that have been implemented in schools in the last thirty years. Where did the idea originate? In the 1960s, New York State instituted a 1,200-minute science lab requirement. In order to fulfill the state requirement, students originally had to stay after school for science labs then take the late bus home.

The after-school classes, however, had a significant impact on the physical education department's ability to schedule games. Most schools, therefore, morphed into providing double periods (90 minutes) for students to complete science lab investigations. Science teachers welcomed the 90-minute double lab periods because students could set up experiments, take data, and then have time to discuss the outcome of the labs. The students love science because they are able to get out of their seats to do experiments.

The 90-minute periods provided more time for students to delve deeper into the curriculum. The idea of double period blocks caught on. What a great idea! Many schools attempted to implement a block schedule with a focus on more time for math. But many schools also reverted to the 45-minute period due to the fact that teachers (other than science teachers, who welcomed the extra time) were not trained in how to deliver instruction in a block.

For math teachers, double periods are an anathema. Schools that institute double periods in math and don't thoroughly provide teacher training wind up with a real torture chamber for students. To make matters worse, some districts schedule double periods for low performing students. Students who do not like math to begin with now have to endure double periods.

Some math teachers adapted to the 90-minute period by presenting two lessons—double the material which, even for advanced students, is a stretch.

When the double period fails, the blame is on the students not being able to endure the work—too much stress; yet in science the students enjoy the additional time. Science teachers make their courses interesting to learn, they engage the students, and only lecture for, at most, fifteen minutes before the students engage in an investigation. These are science teachers who have the opportunity to develop their instructional expertise that leans toward a student-centered collaborative classroom.

Similarly, a standards-based math program like *MathScape* (formerly *MathLand*) engages students with investigations, manipulatives, and challenging problems. However, math exploratory and modeling lessons need more than forty-five minutes to be efficiently implemented. The perfect storm in the case of the double periods is not training the math teachers and then expecting students to initiate investigations on their own.

In summary, training teachers during the school year is like fixing a plane in flight. There is very little that can be done to implement newly learned methodologies into the established routine. The instructional plane is in flight; math teachers have 150 students to teach each day, and they need to administer assessments and grade papers. Their routine is disturbed if they need to write curriculum, also; even if they have resources readily available that just need to be reviewed before they are launched in the classroom. Remember that teachers have no secretaries to help them duplicate materials and no typists to help them develop documents.

What is the answer to training teachers? How do we prepare them to "teach in the block" and be able to manage a class for ninety minutes? Having more time for math, teachers would be able to provide investigations that would engage their students in math. The time to train teachers to teach in a block is not during the school year, however.

Hence the title of this book, *Navigating MathLand*, encompasses the "Land of Math reform" (that is, the continual attempts to bring math education into the twenty-first century from a traditional venue that is 180 years old). Like the perpetual motion of a swing going back and forth, American educators experience short visits to new-age instructional programs, then fall back to the traditional mediocrity. America still ranks between twenty-third and thirty-fourth, depending on the statistical slant, among the nations of the global society. What is needed for the nation to remain in *MathLand*?

In order to effect change in the system, teachers need to be properly trained in how to deliver standards-based curricula. First and foremost, the teachers need to be trained in depth. The next step is to work with teachers as instructional specialist to design and develop a twenty-first-century education system model that is relevant to the students. That means that schools with over 70 percent of their students failing will create instruction to address the educational needs of those students. And while the programs for the failing schools will differ from those for schools where the majority of students enter prepared to learn, the math curricula will be rigorous and engaging for all.

FIVE

Understanding the Implications of Common Core Standards-Based Programs Accepted in American Culture

How Will Your Child's Math Education Be Influenced by the Common Core Mathematics Standards?

THE COMMON CORE EXPLAINED

What is the Common Core all about? It is important for parents to understand how standards-based instruction differs from the traditional content-based approach for teaching math. Chapter 5 will help parents gain an understanding of how standards-based instructional programs (for example, *Investigations, MathLand, Core-Plus*) prepare students to engage in learning mathematics.

The reader will gain insight into what standards-based instruction looks like. Parents will learn that there are consequences from improper implementation of the Common Core and also be made aware of the impact of the pendulum swing from the belief that Common Core is not improving student achievement.

Navigating MathLand is about what rigorous, engaging math lessons look like. Chapter 5 will help parents decipher if their child's math program is addressing all four math learning styles. The intent is not to overwhelm the reader with the mathematics problems or teaching jargon and methodologies but to model the difference in instruction.

The two approaches to delivering math instruction, standards-based and traditionally taught lessons, will be compared. The explanation of each teaching approach will be linked to the four math learning styles inherent in each student, and the eight math practices that render students mathematically proficient. The reader will learn that the most important math practice (arguing a solution to a problem), as well as critical thinking, has been left out of the traditional approach of teaching mathematics.

A day does not go by without a media report on the Common Core. Along with the media dissent is the conundrum that while twenty-first-century teachers are rated "highly effective," their students are not passing the Common Core state math assessments. "How math is taught" has been debated nationally over the past five decades and with all the noise made about math reform, US secondary math students are still ranked below thirty among the industrial nations.

The twenty-first-century standards-based curricula was mandated to be implemented throughout the nation's schools by 2014 and as yet has not had a chance to play out. Three years of test scores is not enough to reveal the effect of the Common Core taught in the elementary grades on students entering middle school. Give it some time.

Massachusetts has implemented standards-based math reform for the past twenty-five years and now the state is edging up in rank among the industrial nations. For that reason it was second nature for the "Bay" state to adopt the Common Core State Standards for Mathematics, the national standards-based program.

There now is a movement by states toward reverting to a more traditional, less rigorous math program. This movement is based on myths and misinformation about the intent of having engaging and rigorous math instruction delivered to all students. The question is, "How will yet another pendulum swing of the 'math wars' affect your child's learning of mathematics now and in the future?" Why should parents be concerned about the inception or demise of the Common Core instructional initiatives?

Some states are rethinking the Common Core entirely, and so did the 2016 presidential campaigns. New York State has always been the leader in state assessments, and its uniform math curriculum requires students to pass a Regents exam in June to obtain a Regents high school diploma. The swing back to the educational drawing board will create another perfect storm, putting all students in math limbo.

In 2016, the educational culture in New York State shifted from progressing with Common Core to going back to the drawing board and revising the current New York State P–12 Common Core Learning Standards for Mathematics document. This regression is a result of teacher evaluations being partially based on poor student scores on mandated state assessments, inadequate training of the teachers leading to instructional misconceptions, and the archaic school schedule that does not permit time for the proper rollout of a new and challenging math curriculum.

Currently, the revisions are posted and awaiting approval by the NYSED, where, for over 100 years, teachers in New York State have prided themselves on the rigor of the Regents program. In the past, the school community informally evaluated teachers based on how their students scored on the Regents exams. Parents and administrators knew who the best teachers were based on the Regents results. It was a badge of honor to teach a Regents course and to have students achieve a 100-percent passing rate.

For some reason, Regents teachers were deemed brilliant and non-Regents teachers were not considered smart. The irony is that it takes a very intelligent, skilled teacher to differentiate instruction for the at-risk, non-Regents student population.

There is an imbalance in the expertise of the teachers teaching the at-risk students and those teaching honors students. The honors teachers always get the brightest and the best, and the at-risk students are assigned to teachers who are deemed not capable of teaching brilliant students. Mr. Kasal was the exception.

The "too soon" pendulum swing raises the questions: (1) What does excellent math instruction look like?; (2) What is the educational classroom climate needed to engage students in learning a rigorous math program?; (3) Why were some schools (charter schools) able to meet with success in teaching Common Core math?; (4) Has the agreed-upon meaning for math literacy changed?; and (5) What can I do as a parent to ensure that math instruction is engaging and rigorous in preparing my child for the college and career demands of the twenty-first century, regardless of the pendulum swing?

In 2011, New York State–certified math teachers were invited from all parts of the state to take part in writing the NYSED EngageNY math curriculum for pre-K–12 grade levels, based on the New York State P–12 Common Core Learning Standards for Mathematics. A standards-based document is not a written curriculum but rather a guide as to what student should know and be able to do at the end of each grade level. It was up to each state to write a curriculum that addressed the Common Core standards. By 2012, the Common Core State Standards for Mathematics, the national standards-based document, was adopted by forty-two states.

The EngageNY math documents were organized into divisions, called modules. After the EngageNY curriculum was completed, each teaching module was posted on the NYSED EngageNY website for all math teachers to review and submit suggested revisions. Each grade level/high school course was divided into four to eight modules. The Common Core curriculum is not perfect, however, and has its own gaps (for example, eliminating the study of logic). Educators have the opportunity to develop and add topics such as logic to the school math curriculum.

The truth is that teachers have been involved in writing standards-based math curricula for over six decades. Teachers who wrote these engaging math lessons were members of the cadre of educators who engaged all students and inspired them to learn mathematics. These were teachers who were well-versed in mathematics, provided rigorous instruction in their classrooms, and were cognizant of the research-based findings on how students best learn math.

Great teachers made learning mathematics fun, interesting, and applicable to a student's life. These teachers realized that the "old" eighteenth-century math curriculum, drill and kill, may have worked then in the nineteenth century and perhaps the early part of the twentieth century, but it does not work now. These are the teachers who realize that teaching is a profession, not a job.

Professionals need to keep up with the latest research. Professionally minded teachers attend math conferences to learn more about research and new instructional strategies, and they also share their successes in the classroom with other educators. As parents, it is important to ask administrators how many math teachers attend math conferences. How many math teachers currently present workshops at conferences and for their district? Turnkey training (teachers teaching teachers) is one of the most efficient ways to propagate newly learned instructional strategies.

All standards-based math programs (curricula), like the Common Core, provide rigorous, engaging instruction for all students, ranging from those who are struggling with math to those who are genius. There is a misconception that the traditional "drill and kill" practice has been lost but, in fact, practice is imbedded in the Common Core program in the form of "sprints" to accompany the lessons. The baby has not been thrown out with the bath water—yes, students still need to be fluent in the multiplication tables!

The shift to standards-based learning is more difficult for students because now the onus is on the student, not the teacher, to problem solve. Students are still emerging from a teacher-centered class. In a traditional setting, it is easier for students to sit at their desks and be told how to calculate an answer to a long division problem with the long division rules (algorithm) than it is in a standards-based program which calls for them to build an understanding as to why the algorithm works.

In the future, students who understand why the long division algorithm works will have a clearer understanding of how to divide algebraic expressions. Standards-based instruction forces students to visualize a problem and to think critically to find the solution. The standards-based lesson does not stop at the students finding a solution to a problem. There is a very important final step to the lessons—all students are required to present their solution to their peers and discuss alternate and more efficient solutions.

This collaboration process fosters understanding of math concepts and skills. More importantly, students have a better understanding of how to apply the concepts and skills to nonroutine problems. A student learns to become confident in solving nonroutine problems, as most problems in the real world are nonroutine. Standards-based problem solving teaches students the process of using concrete models and virtual illustrations to apply the proper algorithm to find a solution. Collaboration brought about by problem solving fosters math concepts and skills.

The big hurdle is how deeply the school community has embraced the Common Core. Many educators believe that the Common Core shall pass; that is, school math programs will visit the Common Core just like educators visited *MathLand*. However, the visit never leads to permanently staying with a rigorous math program. There is the swing back and forth with curriculum. As the reader learned in chapter 4, New York State designed and implemented four math curricula in thirteen years.

The negative sentiment about the Common Core by educators pours out into the parent community. The issue of providing excellent math education through a reformed curriculum often becomes a political football. Meanwhile, the students remain squeezed between a multitude of changed math curricula with few opportunities to connect the math concepts and skills, and without learning to solve nonroutine twenty-first-century math problems.

The question is, "Will the American education system stay with an engaging curriculum, or will there be a return to nineteenth-century traditional/procedural instruction again?" Chapter 4 focused on the American school system industrial model as being structurally poor in supporting standards-based reform. The existing nineteenth-century education structure for secondary schools disconnects math from other major content areas (science, social studies, English) as well as art, music, and physical education.

Parents do not have much control over the length of the school year, day, class periods, or the training of teachers. However, being an informed client of the public education system helps; parents need to be aware of how the infrastructure of the educational system impacts, and often clouds, the reform necessary to promote a "Wright Brothers-Bill Gates-Steve Jobs" perseverance of learning. Parents who are informed can urge that their boards of education provide the instructional environment needed to prepare their children for the twenty-first century.

LEARNING ABOUT STANDARDS-BASED INSTRUCTIONAL PROGRAMS

Parents do need to understand that there are advantages in a standards-based curriculum in America's education system. Even though they were taught with traditional methods, parents need to be aware of the misconceptions attached to the methodologies purported by both the older, traditional way math was taught and the new, standards-based approach. Knowing that the cementing factor for an engaging math curriculum is instruction that addresses the four math learning styles and the eight math practices will make students mathematically proficient.

Walk through any school and observe students in a math class. At the elementary level you will observe students grouped and talking about their math work. In middle school you will see students talking in math class at a lesser rate. In high school you will find, for the most part, that the room is quiet, the teacher is talking, or lecturing; and the students are usually sitting in rows and working individually on worksheets.

Occasionally, the teacher may walk around the room to look at students' work and make comments. Most of the time, when students are asked for the answer to a problem and the correct answer is given, the teacher assumes from the silence that the students understand the solution.

Silence does not indicate that students are engaged in the work or have had their questions answered. The daily teaching routine finds many students hiding in the silence. When a test or quiz comes at the end of the week, students do not find out how they scored until early the following week. By then the students have forgotten questions they missed that they never had answered by the teacher. Feedback is not immediate, especially on daily homework or over a weekend or vacation.

ALERT: Parents, proceed with caution when reading this section. You may feel overwhelmed by the examples rendered. You are not expected to know the answers or understand the strategy, but note that the way you were taught math is not the same way standards-based programs provide instruction. *Navigating MathLand* is a "what is this?" not a "how to" book, and is intended to provide you with a tool to be able to identify engaging math instruction. At the end of the book there will be a resource list if parents wish to review how standards-based lessons are presented.

Subscribing to the adage, "When you know what . . . then you know how," parents can be proactive by asking the board, administration, or teachers if there has been ongoing training for teachers on how to use standards-based programs. By knowing how to identify engaging math instruction, parents do not have to accept claims by educators that the teachers have been trained in the Common Core.

Parents will know that the training that needs to be taught first is about the difference between "content-based" instruction (traditional) and "process-based" instruction (standards-based); and that proper Common Core training is ongoing and collaborative in nature. Parents will also know that first-year math teachers, even if teacher prep programs tout Common Core training, are mostly immersed in traditional school culture, leaving new teachers with idealistic views of creating engaging lessons and experiencing frustration like a salmon swimming upstream.

Basically, most teachers have been handed manuals for standards-based math programs with no explanation of how the program differs from a traditional approach (content-based). If a district decides to adopt a standards-based program and fails to properly train teachers as to how to implement the program using different instructional methodologies, teachers tend to interpret the standards-based curriculum from the "content" lens of instruction with which they are familiar.

Even if math teachers were trained in some standards-based strategies, most of the time the traditional instructional structure of the school inhibits teachers from implementing or shifting to standards-based lessons. Teacher training should be akin to a doctor's internship and residency, requiring a two- to four-year commitment preparatory to actual practice. In addition, there should be a commitment to a profession that requires continued study of instructional methodologies based on valid educational research.

Currently, many math teachers believe they are teaching the Common Core but they are really focusing on the content (fractions, decimals, numeric expression, equations) aspects of the curriculum rather than the process (analyze, synthesize, create, explain, argue). Content-based lessons are, for the most part, teacher-centered, where the teacher has to explain how to proceed with a problem, as compared to standards-based lessons where the teacher is the facilitator of student-centered lessons.

As a result, the content-based approach to math instruction becomes procedural (step-by-step or conceptual), perhaps "hooking" the students with dominant "Mastery" and "Understanding" learning styles, and thus leaving the "Interpersonal" and "Self-Expressive" math learning styles out of the classroom interaction. The short class periods, coupled with content-based, teacher-centered lessons, usually leave no time for students to share their solutions and mathematical thinking with their classmates.

Content-based (traditional) lessons may be rooted in computational procedures with fractions, decimals, and percents. The same lesson can be taught as standards-based using the same content and skills but also taking students through the process of connecting fractions, decimals, and percents to a real-world problem.

The bottom line is, "What constitutes *engaging* math instruction?" The answer lies in how math concepts and skills are introduced to "hook" students into learning. Standards-based instruction is more comprehensive in provoking students to "think" on their own. Elementary school math lessons "hook" students in learning, where more time is set aside for math and the teachers are instructionally savvy.

The issues with delivering engaging lessons begin in middle school. However, grades pre-K–12 Common Core math lessons all employ the eight practices necessary to become proficient in math. Whether students are in a traditional or a standards-based program, educators agree on the importance of the eight math practices as a format for the study of mathematics.

The eight practices make sense and embrace each math learning style. As teachers craft a math lesson, they need to be aware of which math practices are addressed for the day. To do all math practices in one lesson is not prudent, but all practices need to be addressed over a period of the school year. Both the content-based and standards-based lesson philosophies agree on the eight practices.

The eight mathematical practices (MPs) are listed. Mathematically proficient students: (1) Make sense of problems and persevere in solving them; (2) Reason abstractly and quantitatively; (3) Construct viable arguments and critique the reasoning of others; (4) Model with mathematics; (5) Use appropriate tools strategically; (6) Attend to precision; (7) Look at and make sense of structure; and (8) Look for and express regularity in repeated reasoning. The eight practices are defined in depth in the New York State P–12 Common Core Learning Standards for Mathematics document.

TRADITIONAL VERSUS STANDARDS-BASED (COMMON CORE) LESSONS

Along with integrating the eight math practices into both traditional and standards-based lessons, parents need to be able to understand the differences in the lessons implemented in the classroom. The following generic lesson can be taught at the secondary level, grades 6–12. The content of the lesson can be altered depending on the grade level and ability of students. The lesson was specifically developed to provide a stark comparison between the traditional method of lesson design and the standards-based, Common Core approach to math instruction.

The content of the sample lesson is the same for both examples and can be executed in grade 7. The Common Core math curriculum for grade 7 is focused on proportional reasoning. In order to execute the lesson, it is understood that students in grade 7 learn how fractions, decimals, and percents are connected, as well as how to calculate the area of a circle.

TRADITIONAL LESSON

The teacher plans to review the multiplication of fractions with the class and then have the students represent the products (answers to multiplication) as decimals and percents. The "aim" for the lesson (to learn how to change fractions to decimals and percents) is written on the board. The bell has rung; the period 2 math students file out of the class. In four minutes the classroom is filled with period 3 students. The late bell rings, and the thirty students are seated in five rows, with six seats in each row.

A typical secondary (middle or high school) 45-minute traditional math lesson is presented as follows:

Step 1—As the students enter the classroom, a "Do Now" problem is written on the board. This may be a problem relating to yesterday's lesson. The teacher begins the class by asking the students to complete the following problems on the board—that is, to find the product for the following problems: 1/3 x 1/7; 2/5 x 5/8; 3/5 x 15/22. Students complete the "Do Now" and copy the aim in their notebooks as the teacher takes attendance and checks homework.

At this time, some students may be asked by the teacher to go to the board to write their solution to a homework problem. Usually the teacher selects the students who have the correct answers go to the board. The beginning of the lesson may take five minutes.

Step 2—The teacher reviews the "Do Now" and the homework with the students. The correct answers to both the "Do Now" and the homework are on the board. The teacher asks if there are any questions. There may be one or two from the students who have completed the homework and have questions. The students who did not do the homework are left out of the review. Rarely does the teacher ask students to put up the solution to the same problem so that alternate solutions could be addressed. This part of the lesson may take fifteen to twenty minutes.

Step 3—The teaching portion of the lesson. The teacher introduces a new concept or skill. Here, the teacher demonstrates how to do the new work by the step-by-step approach, using an algorithm. For example, if students learned how to multiply fractions yesterday, today they will learn how to convert fractional answers to decimals and percents.

All the students are required to take notes on how to change fractions into decimals and percents. As mentioned in the previous chapter, most students are busy copying the notes written on the board and not given a chance to comprehend what they are writing. The teacher usually continues to write more notes. Or the teacher provides fill-in-the-blank "guided notes," which cuts down on the need to copy entire sentences but still does not give the student time to reflect on the phrases.

Students are expected to use the class notes to complete a homework assignment for tomorrow. Scientific calculators may be distributed to facilitate or check the calculations so students will not have to do long division. This may take fifteen minutes. The teacher explains how to do the problem and get the correct answer.

Step 4—Students complete practice questions. If time (five minutes) permits, students practice the "new work" using similar problems posted by the teacher. The teacher circulates throughout the class and stops to help students who appear to be struggling. Students are asked to copy each sample problem and solutions in their notebooks. The teacher hands out a worksheet as the homework assignment (it contains twenty to thirty similar problems that were modeled for the students in the class lesson).

If the student does not understand the day's lesson, there may be no support (except mom or dad or friends) in the afternoon to help with, do, or copy the homework. The student has to wait until tomorrow's lesson to ask questions and get feedback.

The notes the student has copied from the board do not help with the homework because the student never had time to discuss and interpret his or her notes to solve problems. The notes are not meaningful because he or she copied the notes without thinking about what they meant. Tomorrow the class moves on to the next lesson (that is based on an understanding of today's lesson), leaving the confused student behind. The teacher offers help twice a week after school and the student can opt to go for help.

Friday is usually quiz day. Math teachers give quizzes on Fridays, not Mondays, for fear that students will forget what they learned over the weekend. The quizzes are collected and graded; and perhaps the students will get the results on Monday. There is a time lapse between the student arriving at a solution and knowing if the answer was correct. Timely feedback to students while they are in the process of learning the concept or skill is not available. Students need time to reflect on where there was a mistake.

If the teacher has an "Instrumentalist" philosophy, students will need to show mastery in the skill before they are asked to apply the skill to a real-world problem. If students never reach the skill level to the teacher's satisfaction, the teacher will have them continue to work only on skills in the lessons and never get to the application. This "drill and kill" approach to math instruction is usually found in classes with struggling students and students with special needs. It's like practicing to play a sport, but never getting a chance to play a game.

The traditional lesson is content-driven with the goal of familiarizing the student with a particular procedure for solving a problem. Teachers feel that they must "tell" students how to proceed to solve a problem using an algorithm. As a result, students may suffer through many lessons, such as those on long division (how to divide step-by-step), scaffolding up to advanced problems such as dividing a five-digit number (25,463) by a three-digit number (345) and writing the remainder as a fraction or decimal (the answer, 73.8057971014, rounded off to the nearest tenth is 73.8).

Traditional lessons were intended to connect fractions with decimal equivalents and then have students represent the decimal as a percent. The same skills and content (fractions, decimals, percents) can be integrated into standards-based lessons. For the sample standards-based lesson, the teacher uses standards-based strategies to introduce students to the concepts and skills needed to complete the lesson. However, the standards-based lesson is designed for students to *apply* their understanding of multiplying fractions, using decimal equivalents and percents to solve a problem.

STANDARDS-BASED LESSON

Parents have a right to visit school math classrooms and to explore how the math curriculum is taught. By visiting all middle school math classes, parents will be able to witness the gamut of traditional to standards-based lessons. The importance of the visit is to determine if the lessons are engaging. Upon visiting the math classrooms, parents might experience a standards-based lesson being taught as follows:

Step 1—The teacher launches the lesson (ten minutes). The "launch" usually contains a "hook," information needed to do the lesson and solve the problem. The teacher has written on the board the day's goal, "I will understand how to create a portion size unit for a pizza pie." Students are expected to copy the goal in their notebooks.

In front of the students are two rectangles (a large rectangle 8" by 12" and a smaller one 8" by 9"). Students have access to rulers. The "Do Now" written on the board is a set of instructions asking students to represent the answer to the multiplication problem 1/2 x 1/4 using the large rectangle, and the answer to the multiplication problem 1/2 x 1/3 using the smaller rectangle. The students have learned how to illustrate fraction multiplication using the area model strategy.

Once the students calculate and illustrate each answer to the multiplication problem, the teacher asks the students to compute the area sizes of the answers. The students will find that the unit representing the answer 1/6 for the smaller rectangle and 1/8 for the larger rectangle is the same size: 4" by 3" (both are 12 square inches). However, depending on how the students grid the paper, there may be a variety of correct rectangles—12 square inches—with different dimensions, such as 4 1/2" x 2 2/3" for the small rectangle and 2" x 6" for the larger rectangle.

The point is that all students are actively engaged in representing multiplication of fractions. The task has the students create visual representations of the computations, with the goal of establishing a unit of portion size. It also allows time for the teacher to assist students who need help, or to invoke students to instruct each other.

The "Do Now" rectangles completed by the students will be used as manipulatives for today's lesson. The teacher walks around the classroom to make sure that every student is on task and that students are using the rulers to measure the dimensions in inches. The students are recording the process in their notebooks using pictures and text to describe their answers.

The students have already completed the previous night's homework—multiplying fractions using the area model strategy as well as the algorithm on a web-based math program. The website provides instant feedback, giving the students answers to the homework questions as well as explaining wrong answers. Students paraphrase notes from the web-

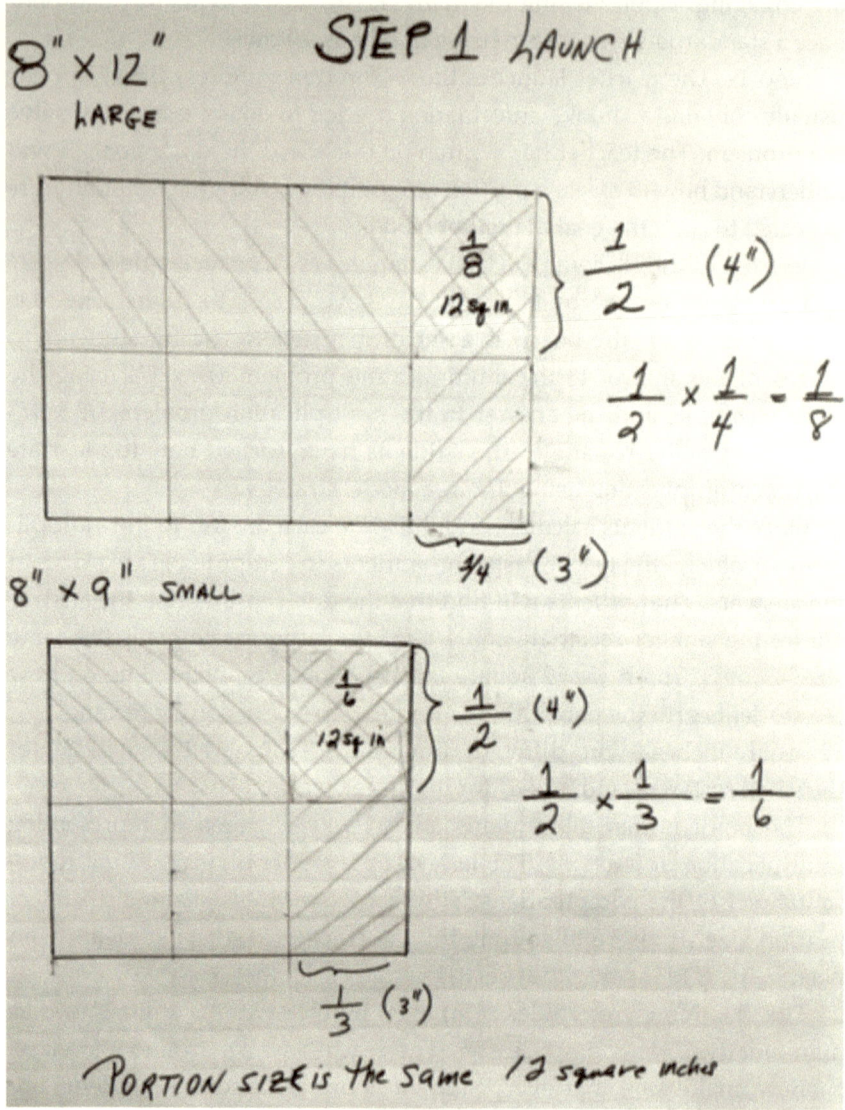

based homework and record in their journals any questions about the process. The results of the homework are sent to the teacher who analyzes the data in the morning to help guide the day's lesson.

The teacher then launches the lesson. Today students will investigate the question, "Are large pizzas cut in eighths and small pizzas in sixths to create equal portion sizes?" The students are to use the rectangular ma-

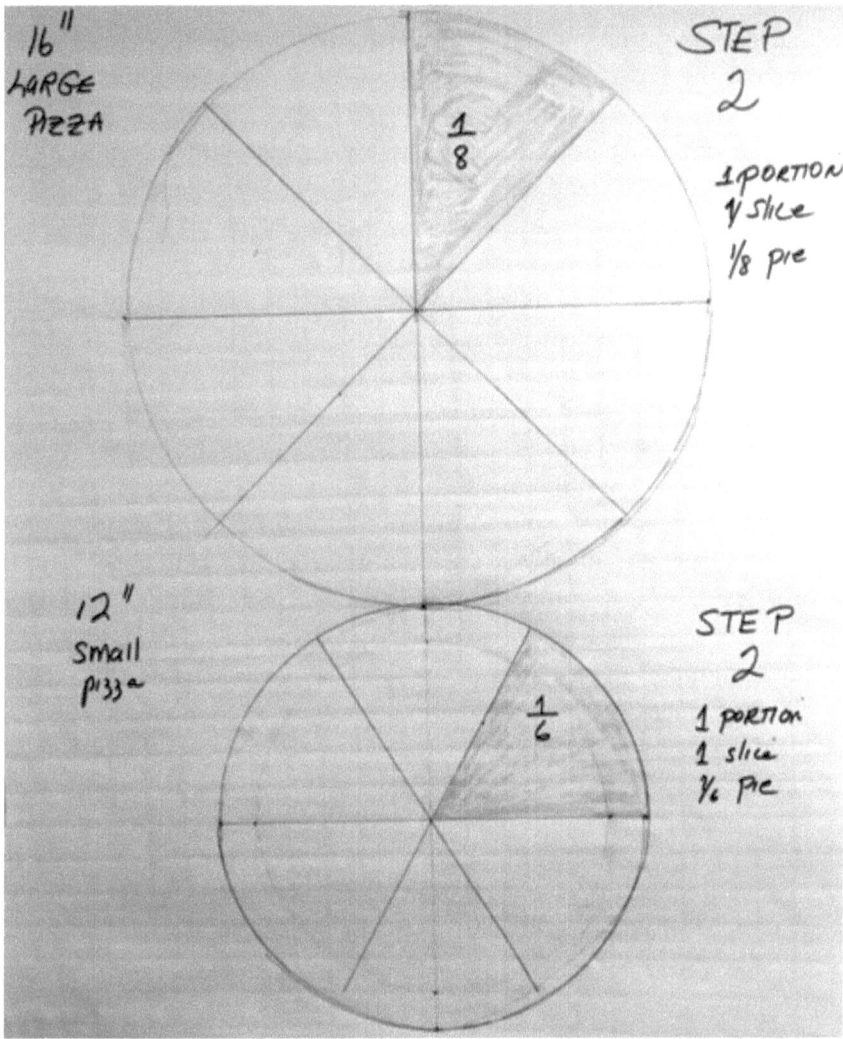

nipulatives they used as a "Do Now" to help them address portion size. The large rectangle is cut into eight pieces and the small rectangle is cut into six pieces. Both answers to the fraction multiplication rendered a same-size piece of the rectangle, 12 square inches.

Prior to today's lesson, the teacher used the feedback she or he received for each student's homework to group the students for the lesson task. The students are grouped in twos, threes, or fours, based on the teacher's discretion and understanding of students' abilities/learning styles. The teacher also has the option of having the students work individually on the problem first, then in groups. The teacher, as a professional, has the expertise to develop groups to provide the optimum peer collaboration.

The teacher may introduce the task by saying, "I'm sure you have ordered large pizzas and small pizzas. Did you ever wonder why the large pies are cut into eight pieces and the small pies are sliced into six pieces? How do the portion sizes compare?"

The students will use the next twenty minutes to create explanations for the following problem: "Provide the mathematical rationale for why large pizzas (16" in diameter) are cut into eight equal pieces and small pizzas (12" in diameter) are cut into six pieces. Are large pizzas cut in eighths and small pizzas in sixths to create equal portion sizes?" Use today's "Do Now" findings to guide your answer.

Step 2—This is the "explore" portion of the lesson (twenty minutes). Students work individually or in small groups (they get "messy" with math). In grades 6–8, students are permitted to use scientific calculators to find answers in order to expedite the many calculations that are needed to support their solutions.

Students must first discuss a plan on how they are going to solve a problem based on an estimated answer. The goal is to get the students to realize that the slice of pizza for 1/8 of the big pizzas may be the same size as the 1/6 slice of the smaller pizza. Or is it?

Each group is expected to report their answer to the class during the last five minutes of the explore session. The teacher has given each group a large sheet of poster paper to present their answer. Students also are given compasses and protractors to draw the exact size of the pizza and divide the pizza into equal slices. They also, as a group, use the participation rubric to evaluate their participation in finding the solution as well as how the group worked together.

Step 3—The summary (ten minutes) is where the main teaching occurs. Using the results found by the students, the teacher goes back to the goal of the lesson and asks the students to explain what is meant by portion size—a unit derived by a restaurant. The discussion is around the size of the portions and may include a query about what if pizzas were rectangular—like the manipulatives?

Are the portion sizes close in size for the large and small pizza? When a student orders a slice of pizza, is it taken from a large or small pie?

Students will have determined that the large size slice is 4/3, 1.333..., or 133.33 percent larger than the small size. In the process, students will need to use decimals and percents to explain their answers.

Step 4—Finally, there is an assessment of the lesson. The teacher distributes an exit ticket, a quick question that students answer before leaving class. This provides feedback for the teacher as to whether the students understood portion size. Are the portion sizes the same for a large pizza and a small pizza? How does changing the shape of pizzas to rectangles ensure that the portion sizes are equal?

This example of a standards-based lesson is process-driven with the goal of understanding a concept or skill to which students must apply the concept of portion size. The homework for each lesson differs as follows:

CONTENT-BASED HOMEWORK

First problem: Multiply 1/2 x 1/3 then represent the product as a percent. The student multiplies numerators 1 x 1 = 1 and multiplies denominators 2 x 3 = 6. The answer is 1/6. The student then divides 1 by 6 which equals .1666..., a repeating decimal; then multiplies the answer by 100 to equal 16.6666...%; then rounds the decimal to the nearest tenth to 16.7%. The answer is 16.7%. The worksheet asks students to compute answers to twenty similar problems involving fraction multiplication, to decimal, and to a percent.

COMMON CORE STANDARDS-BASED HOMEWORK

For homework, use the large (16") and small (12") pizza sizes from today's lesson or visit a pizza parlor online. Or you may ask for the price and size of a large and small pizza from a local pizzeria. Calculate the portion size for each pizza and the cost per square inch for each pizza size. Compare the price of a portion size of the small and large pizzas. When would it be beneficial to order a small or large pizza pie?

COMPARING STANDARDS-BASED TO TRADITIONAL-BASED

The traditional (content-based) lesson uses primarily passive teaching methods, while the standards-based lesson uses primarily participatory teaching methods. All students are engaged in the standards-based lesson, providing the teacher with rich feedback for each student. The pizza pie problem engaged all learning styles—"Mastery," (the calculations); "Understanding" (developing a plan to understand the process used to determine portion size); "Self-Expressive" (the opportunity to develop the visuals for the solution); and "Interpersonal" (being able to apply the lesson to the real world).

This raises the question, "What is needed to shift teaching from content-based to standards-based?" Regardless of whether or not teachers adopt the Common Core, good math instruction engages students.

There are three shifts needed to occur that are most essential to creating a standards-based Common Core instruction: (1) Focus—Focus deeply on only the concepts that are prioritized in the standards; (2) Coherence—Educators carefully connect the learning within and across grades where students build new understandings onto foundations built in previous years; and (3) Rigor—Pursue with equal intensity the three aspects of rigor as a major work of each grade: conceptual understanding, procedural skill and fluency, and applications.

Comparison of the two approaches to developing math instruction will focus on rigor. Table 5.1 compares and contrasts the two instructional approaches, traditional and standards-based. The question is, "How do we shift from one to the other?"

If a district decides to adopt a standards-based program, it must properly train teachers how to implement the program. The three-pronged nature (fluency, conceptual understanding, and application) of rigor undergirds a main theme of the Common Core with equal intensity to drive instruction for students to meet the standards' rigorous expectations for each grade level.

EngageNY, *A Story of Units*, defines the three-pronged nature of rigor as:

> Fluency—Students are expected to have speed and accuracy with simple calculations; teachers structure class time and/or homework time for students to memorize, through repetition, core functions.

Fluency represents a major part of the instructional vision. At the elementary level it is a daily, substantial, and sustained activity. One or two fluencies are required for each grade level and fluency suggestions are included in most lessons. Implementation of effective fluency practice is supported by the lesson structure.

Understanding the Implications of Common Core Standards-Based Programs 113

Table 5.1.

Rating Category	Standards-Based (Common Core) Pizza Lesson	Traditional Fraction Lesson
Teacher/Student Focus	Student-centered—Tasks are designed from the beginning of the lesson to engage the student in thinking and demonstrating how to model fraction multiplication. Students "discover" that the different size rectangle has produced the same size area. The students use class time to discuss the portion size of pizzas applying fractions, decimals, and percent.	Teacher-centered— "One size fits all"—It is assumed that all the students know how to multiply fractions and are ready to learn the steps to write the fractions as decimals and percents.
Math Learning Styles	All learning styles are addressed. "Mastery" and "Understanding" follow procedures to understand same portion size. There is a "hook" for the "Interpersonal" style to discuss the portion size relative to the size of the pizzas. The "Self-Expressive" student has the opportunity to explore the portion size for a large pizza versus a small pizza, a nonroutine problem.	"Mastery" is addressed. "Understanding," maybe! There is no "hook" for "Self-Expressive" or "Interpersonal."
Math Practices	Math Practices 1, 2, 3, 4, 5, & 6	Math Practices 5 & 6
Rigor	Fluency Conceptual Understanding Application	Fluency
Independent Work	Homework is meaningful and practical.	Homework is rote with repeated questions.

Fluency tasks are strategically designed for the teacher to easily administer and assess. A variety of suggestions for fluency activities—including mental math activities, interactive drills, quick and efficient games with dice, spinners, and cards, and concept worksheets—are offered. Throughout the school year, such activities can be used with new material to strengthen skills and enable students to see their accuracy and speed increase measurably each day.

> Conceptual Understanding—Students deeply understand and can operate easily within a math concept before moving on. They learn more than the trick to get the answer right. They learn the math.

Conceptual understanding requires far more than performing discrete and often disjointed procedures to determine an answer. Students must not only learn mathematical content, they must also be able to access that knowledge from numerous vantage points and communicate about the process.

A standards-based lesson is designed for students to use writing and speaking to solve mathematical problems, reflect on their learning, and analyze their thinking. The lessons and homework require students to write their solutions to word problems several times a week.

Thus, students learn to express their understanding of concepts and articulate their thought processes through writing. Similarly, students participate in daily debriefs and learn to verbalize the patterns and connections between the current lesson and their previous learning, in addition to listening to and debating their peers' perspectives. The goal is to interweave the learning of new concepts with reflection time into students' everyday math experience.

> Application—Students are expected to use math and choose the appropriate concept for application even when they are not prompted to do so.

Standards-based Common Core lessons are designed to help students understand how to choose and apply mathematics concepts to solve problems. The EngageNY curriculum includes mathematical tools and diagrams that aid problem solving, interesting problems that encourage students to think quantitatively and creatively, and opportunities to model situations using mathematics.

The goal is for students to come to see mathematics as connected to their environment, to other disciplines, and to the mathematics itself. A range of math problems are presented within the modules, topics, and lessons that serve multiple purposes.

TEACHER BELIEFS AND CONSTRAINTS

The teacher's perception that the standards-based approach works is based on his or her beliefs about how math is learned and how math should be taught. Even if a teacher has intense training in the Common Core or any other standards-based program, if the teacher does not believe in the Common Core approach, their ability to implement a standards-based lesson is compromised.

When a teacher is forced to teach in a style with which he or she is not comfortable, students pick up the negative sentiment, verbally and nonverbally (body language). Engaging instruction is thus thwarted. For example, if a teacher has an "Instrumentalist" philosophy (believes that practicing unconnected skills is what math is), practicing skills is the way math is learned ("drill and kill"). With a "Mastery" math learning style, and a "Mastery" teaching style, it is nearly impossible to have that teacher shift to a standards-based approach. The teacher needs to acknowledge his or her beliefs and then acknowledge a need to change. Change can happen.

Teachers can begin to shift toward the standards-based approach, but first they need to identify their own math learning style, mathematics philosophy, and teaching style before they are introduced to standards-based programs. Teachers are generally open to learning about themselves as instructional experts. One of the least explored areas in training of math teachers is having them reflect on what they believe the study of mathematics is about and how students prefer to learn math.

As a result, most math teachers' lessons reflect how they best learn math. This often results in teacher-centered (not student-centered), content-driven lessons that, at best, "hook" 50 percent of their students. Compounding the ability to shift instructional strategies is the culture of the traditional school setting in which they practice. Even if teachers attempt change, there are always issues beyond their control that impede any shift or reform in their instruction.

School culture often compounds a teacher's ability to provide students with engaging math instruction. These school cultural issues include the socioeconomic level of the school community; student performance levels on past state assessments; the culture of the school; the ideologies of administrators and educators; mathematics department constraints; budget; teacher daily class schedules; class makeup (heterogeneous/homogeneous abilities); district math curriculum decisions; and resources (technology, texts, manipulatives, etc.), to name a few.

In summary, a teacher's professional identity—their system of beliefs concerning mathematics, its teaching, and learning, their level of thought processes and reflection on practice, and the constraints and opportunities provided by the social context of the teaching practice—has a major impact on instructional decisions. Once teachers are aware of their beliefs, learning styles, teaching styles, how they reflect on their practice, and how the culture impedes their autonomy, they can begin to shift.

When training math teachers in the Common Core, the secondary teachers must be apprised of the entire curriculum, grades pre-K–12. Math teacher-training needs to be comprehensive in content and include elementary and secondary training. Elementary instructional strategies can be applied to secondary instruction. Students at all levels need to be able to illustrate a math problem using tape diagrams.

Grade 5. Example using a tape diagram to illustrate and solve a problem (taken from EngageNY, "How to Implement A Story of Units," 2013, p. 43).

Sam has 1,025 animal stickers. He has 3 times as many plant stickers as animal stickers. How many plant stickers does Sam have? How many stickers does Sam have altogether?

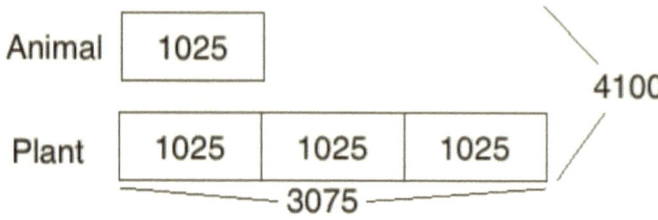

1 unit = 1,025
2 units = 2,050
3 units = 3,075
4 units = 4,100
1. He has __3,075__ plant stickers.
2. He has __4,100__ stickers altogether.

BLENDED CURRICULUM/FLIP CLASSES

Some school districts have a blended curriculum (mixing standards-based and traditional), and some have initiated flip classes (students listen to lectures online as homework, then work on standards-based lessons in class). The difficulty lies in the time element. It's difficult to get both the lecture portion and the "hands-on" portion of the lesson squeezed into one 45-minute period.

To thwart the fear of not being able to provide the "drill and kill" that teachers believe students need to learn math, some districts have created curricula that integrate standards-based with the traditional approach. This starts out as a good idea, but within a few cycles, there is a tendency to revert to the traditional math program.

If teachers are properly trained in the Common Core, they will realize that the standards-approach includes "drill and kill" exercises, now renamed as "fluency." Standards-based is not given a chance, and as soon as the same levels of scores are reported, the district backtracks to the safe traditional program.

ENGAGENY NYSED COMMON CORE LEARNING STANDARDS (CCLS) FOR MATHEMATICS CURRICULUM

Parents should know that EngageNY provides an extremely rigorous standards-based math curriculum written by educators (including teachers) for grades pre-K–12. For each grade level there are teacher materials (instructional manuals and materials) and student materials to accompany the manuals. The grade level curriculum is divided into four to seven modules. Each module is broken into two to five math topics; each topic is divided into lessons.

The EngageNY manuals (modules) are written with in-depth directions for teachers and can have up to 600 pages of information. Generally, the total number of pages for the teachers' modules can average 1,300 pages per grade level! The information on how to teach each EngageNY model lesson is clearly written.

Teachers have the opportunity to read the model lessons and interpret them using their professional expertise. The issue is that teachers often interpret the information as content-based and don't have a clear understanding of the process of how to roll out a standards-based lesson. The teacher approaches the EngageNY lesson from a content perspective and overlooks the process.

For example, a content-focused teacher might instructionally evaluate the pizza lesson in terms of the skills necessary for the lesson, spend time on the skills and the step-by-step procedures to calculate portion size, and overlook the visual representation of the multiplication of fractions.

The teacher will then give the answers to the portion size and tell the students that the slice from a small pizza is not the same size as a slice from the large pizza.

HOW COMMON CORE LESSON FORMATS DIFFER

On the secondary level, beginning in grade 6, there are four types of lesson formats EngageNY provides as models for instruction. The four different approaches to delivering a math lesson are germane to the secondary level (grades 6–12). The EngageNY CCLS for Mathematics teacher instruction manuals identify the four lesson type structures as follows:

- Problem Set (P)—Students and teacher work through examples and complete exercises to develop or reinforce a concept. This lesson structure is closest to the traditional procedural teaching, supporting "Mastery" learning and teaching styles. The difference is that in every lesson there is a set of tasks that encourage students to write about and discuss the concept.
- Socratic (S)—Teacher leads students in a conversation to develop a specific concept or proof. The questions asked of the students are specifically developed to lead students to understanding a concept, supporting "Understanding" learning and teaching styles. Built into the lesson are tasks that encourage students to write and discuss the concept.
- Exploration (E)—Independent or small groups work on a challenging problem followed by debriefing to clarify, expand, or develop math knowledge. The pizza lesson was exploratory, supporting all four learning styles; but in particular, it "hooks" the "Self-Expressive" learner. The teacher with a "Self-Expressive" teaching style is more likely to be comfortable with this approach.
- Modeling Cycle (M)—Students practice all or part of the modeling cycle with real-world or mathematical problems that are ill-defined. The pizza lesson could be taught as a modeling lesson if the students were asked to arrive at a size and shape of pizza that would be profitable. The major emphasis of modeling is the application of math skills and concepts to a real-world situation, answering the question asked by "Interpersonal" learners, "So what?! Why am I learning this math?

Even though the four different lesson formats employ four different approaches, math teachers often interpret standards-based lessons from a content perspective with a procedural approach and roll out the lesson in a traditional way. Oftentimes, teachers will attempt a change in teaching a lesson, and then report that the approach did not work. When this happens the teacher needs to reflect and ask, "Why did this approach not work?" If the lesson is rolled out in a manner contrary to the teacher's belief, the lesson will not be effective.

Besides shifting instruction from content-based to process-based, a second instructional conundrum for math teachers is providing a CPA approach to building math concepts: "C" stands for concrete; "P" stands for pictorial; and "A" stands for an abstract/algorithmic approach. To build skills and concepts, a CPA approach to solving a math problem lies in the nature of the lesson, providing a concrete experience, an opportunity to draw a picture and write about the explanation, and the means to create a mathematical expression or equation representing the process.

We will use the pizza lesson as an example. The basic goal was to provide a concrete example to demonstrate portion size and apply the concept of portion size to large and small pizzas. The following examples are used to illustrate the CPA approach.

> Concrete: The "hands on" gathering of real-time data. For the pizza/portion size problem, the students used the data from rectangles to identify or create the notion of a portion size (a piece of pizza 12 square inches in area for both the large rectangle and small rectangle). Students got to measure concrete examples of rectangles divided into eight and six pieces.
>
> Pictorial: Students then had to construct a large pizza and a small pizza and use pictures to depict portion size as a slice. The pizzas are circular. The students may opt to construct pizzas with a radius of 8" for the large and 6" for the small or use scaled diagrams. They divide the large into eight equal slices and the small into six equal slices. Some students may use a protractor to create central angles of 45 degrees for the large pizza and 60 degrees for the small pizza.
>
> Abstract: Students had to calculate the area of each pizza using a formula for area, and then divide each pizza up into slices to determine if the slices represented the same portion. The end result is a set of calculations that are used to determine the portion size of the large and small in order to compare the sizes.

There is a gray area to the acceptable solutions. The large pizza portion is 1/3 more that the small portion. Can we consider this close? Some students might say that the crust on each pizza might influence portion size so the portions are close. Another group may determine that the portion size is not the same. The ratio of the large pizza portion to the small pizza portion is 4:3. Teachers can assign homework and ask students to find the prices and sizes for large and small pizzas and calculate which is more economical.

MATH PRACTICES—CPA

Basic formats of EngageNY lessons—along with real-world applications—give students the platform to discuss mathematics and argue their solutions and critique others. Students still need: (1) to practice and become fluent in math skills (place value, times tables, adding, subtracting, multiplying, and dividing fractions and decimals, solving percent problems, learning geometric shapes); (2) to be able to read and comprehend math problems; and (3) to apply concepts and skills to solve a real-world problem.

Standards-based instruction provides the rigor needed for students to learn the math content and the processes. The next time students order pizza, they will have not only a mathematical view about portions and pricing comparisons but also an insight into business. Don't be alarmed at papers that come home with concrete and pictorial models. Especially in elementary work, students are provided with categories that indicate the primary application of the area of each model.

In summary, as parents, you will need to know about standards-based instruction and what it looks like. Parents need to know that without the proper training, teachers cannot deliver a standards-based curriculum. Before teachers are trained, they need to explore their beliefs about mathematics teaching and learning. Even with intense training, teachers may have a hard time delivering a complete lesson in one class period. Parents need to know what makes content-based lessons different from process-based lessons.

How to do the problems is not the focus of *MathLand*. The intent is to inform parents about the differences, the benefits, and the feasibility of their child engaging in a standards-based program. Not to worry, standards-based programs like the Common Core also include tasks for students to practice and become fluent in math skills.

SIX
Next Steps

What Can You Do to Improve Your Child's Interest in Learning Mathematics?

Maximo entered middle school in the fall of 2015 and is currently in grade 7 for the 2016–2017 school year. More than three years have passed since Maximo's parents were having trouble helping him with his fourth grade Common Core math assignments. Maximo's mom, Vilma, reported that her son's achievement on the grades 4 and 5 NYSED math assessments remained at level 2, which was the same score for grade 3.

Sadly, Vilma was never provided with any feedback or clinical assessment of Maximo's mathematics progress. She did not know what her son needed in order to improve his understanding of mathematics or what he needed to prepare for a rigorous middle school math program.

There is a happy ending, but not without damaging Maximo's belief that he can learn mathematics. Vilma reported that Maximo was selected to attend the grade 6 accelerated math and science curriculum program by his fifth grade teacher who, in Vilma's words, "engaged" Maximo in learning. However, Vilma decided to opt out her son for the spring 2016 NYSED Math and ELA assessments. Vilma's rationale was that the three years that Maximo remained at level 2 had dampened his embrace of mathematics. Vilma was correct. The state test was of no help.

The advanced school program in which Maximo is enrolled promotes a rigorous math curriculum based on the NYSED Common Core math curriculum (the grades 6, 7, and 8 curricula have been compressed into grades 6 and 7). This compression of the curricula allows eighth grade students in the advanced program to take the Algebra 1 Common Core Mathematics Regents in June 2018.

Prior to his move to the grade 6 middle school accelerated program, the grade 5 teacher needed to assess what gaps might exist in Maximo's skill level. Furthermore, the teacher needed to prescribe exactly what Maximo needed to learn to succeed in the accelerated middle school math program. Maximo never learned what math skills he needed to improve or what concepts he excelled at understanding and applying.

According to NYSED, a student who achieves a level 2 on a state exam has gaps that need to be addressed in order to experience success in a math grade 6 program. A question that needs to be researched is, "Are students who underperform on Common Core state tests able to meet with success in accelerated math programs?" If so, are we preventing students from attempting to take more rigorous math programs due to a test that cannot correctly identify the math achievement levels of students?

In the past, when there were only high school state assessments and SATs, high school teachers were adept at evaluating students' knowledge and skills and were able to fill their learning gaps. Maximo will survive and will pass the Common Core Algebra I exam in June 2018 because he is in a positive learning environment that is both engaging and supportive. His teachers and administrators believe that he can learn. He is with students who are excited about learning and with teachers who hold all students to higher standards.

Who knows if the Common Core Algebra I curriculum, let alone the state exam, will be the same by the time Maximo sits for the 2018 exam? Thus, we have another perfect storm! The pendulum seems to be swinging backward. From 2013 to 2015, students in New York State were not performing on the Common Core math assessments to the expectations of NYSED educators. As a result, New York State has reopened the gates of reform and invited math teachers to review, revise, and rewrite the Common Core mathematics curriculum.

Logic has prevailed and New York State teachers have been exonerated from having their yearly teaching evaluations (APPRs or Annual Professional Performance Reviews) tied to the state assessment (grade level and Regents) scores of their students. They are off the hook. To make things more complicated, the presidential campaign of 2016 had politicians promising to disband the Common Core standards document, calling the Common Core a curriculum, not a set of learning standards. As previously noted, a standards document is not a curriculum.

This "backward reform" is not only happening in New York State but also in other states across the country. Look at the bigger picture. Regardless of adherence to the Common Core, the United States is still way behind other countries as far as being competitive in math. The question, "How does the achievement of American students compare to that of students in other countries?" was answered by the National Center for Educational Statistics:

> The Program for International Student Assessment (PISA), coordinated by the Organisation for Economic Co-operation and Development (OECD), has measured the performance of 15-year-old students in mathematics, science, and reading literacy every 3 years since 2000. In 2012, PISA was administered in 65 countries and education systems, including all 34-member countries of the OECD. In 2015 the US moved from ranking 28th in math to ranking 35th when compared to other nations.

In addition to participating in the US national sample, three states—Connecticut, Florida, and Massachusetts—opted to participate as individual education systems and had separate samples of public schools and public school students included in PISA to obtain state-level results. PISA's 2012 results are reported by average scale score (from 0 to 1,000), as well as by the percentage of students reaching particular proficiency levels.

In 2012, average scores in mathematics literacy ranged from 368 in Peru to 613 in Shanghai, China (CHN). The US average mathematics score (481) was lower than the average for all OECD countries (494). Twenty-nine education systems and two US states had higher average mathematics scores than the US average score, and nine had scores not measurably different from the US score.

The twenty-nine education systems with scores higher than the US average score were Shanghai-CHN, Singapore, Hong Kong-CHN, Chinese Taipei-CHN, the Republic of Korea, Macao-CHN, Japan, Liechtenstein, Switzerland, the Netherlands, Estonia, Finland, Canada, Poland, Belgium, Germany, Vietnam, Austria, Australia, Ireland, Slovenia, Denmark, New Zealand, the Czech Republic, France, the United Kingdom,

Iceland, Latvia, and Luxembourg. Within the United States, Massachusetts (514) and Connecticut (506) had scores higher than the US average.

As mentioned in chapter 5, it is no surprise that Massachusetts has out-scored the rest of the United States. A standards-based math curriculum has been a staple in Massachusetts' education for the past twenty-plus years. Education in Massachusetts has been college- and career-directed since the implementation of the Bay State's standards-based curricula. It has taken a generation, twenty-five years, for reform in curricula to be effective.

The fifteen-year-olds from Massachusetts who took the PISA exam in 2012 had been taught a rigorous math program designed by math educators. In Massachusetts, all students are expected to take algebra by the time they reach grade 8.

However, the rest of the United States had only three years to fully implement the Common Core. Over forty states adopted the Common Core mathematics and ELA curricula in 2011; and some states are now "un-adopting" it. By 2016, the education system's Common Core initiative folded to those opponents whose rationale for dismantling the curricula placed blame on the level of difficulty, rigor, and the federal government.

The postmortem on Common Core programs does not mention the failure of its rollout as due to poor or nonexistent training of the teachers and to the complicated eighteenth-century school schedule—an industrial model that does not provide the proper infrastructure to successfully implement standards-based instruction.

What is most disturbing will be the possible shift back to a traditionally taught math program that will leave teachers waiting for the next curricula edict from high-ranking educators. As a result, parents will be left in limbo and students will be the losers. In particular, students will lose exploratory learning experiences and opportunities to discuss math with other students and argue their solutions to complex problems. A standards-based curriculum cannot fit well into an archaic system. What's a country to do?

The math education pendulum is swinging back. What has been perceived as a failure of the institution of the Common Core is the same perception that has thwarted all standards-based programs. Were there any schools that offered a different approach to education? What would a future school curriculum look like? What is the formula needed to improve the school setting?

CAN AND DO SUCH SCHOOLS EXIST?

A story ensues regarding the Pelham Alternate Cooperative Team (PACT). PACT, the name given to a program that was developed in 1976 in the Pelham School District, Pelham, New York, was developed for students who were not functioning in the general high school program. There were forty-two high school students identified as "at-risk"—in danger of not graduating due to attendance problems. The students were not identified as special needs, making the development of school curricula challenging.

There were two teachers selected to teach the PACT students. One was science/math-certified and the other was certified to teach social studies and ELA. Both teachers were special in that they could set goals for students in a "tough love" manner but were also flexible in understanding the individual issues each student brought to school.

The school was developed around a "caring community" theme. There were school general meetings held once a week, and "human development" groups led by a teacher and a counselor. The general school meetings covered how school rules and day-to-day issues were adjudicated. The human development groups fostered discussion regarding issues that students were experiencing with relationships with parents and friends. The PACT curriculum was based on New York State Regents syllabi for science, math, social studies, and English.

Students developed and signed a contract for each course. Courses consisted of lectures that everyone had to attend, and individual and project work. The school accepted students as sophomore, juniors, and seniors. It was best to have all students give freshman year a shot at traditional instruction before coming to the alternative school. The school was not for everyone—just the students who fit. Students could still take regular classes in the high school, to meet graduation requirements or to explore other interests.

Students were also encouraged to have jobs in the community. There were a bevy of local businesses that would accept students as interns. In the 1970s it was easier to be flexible with the course content. For example, a world history course was developed where history was taught through the lens of art history. PACT welcomed retired teachers to provide extracurricular courses.

Mr. Kasal, now Dr. Kasal, had retired from teaching and volunteered his time to work with students who needed to explore building projects. Students and staff welcomed his expertise. It was impressive to enter the school building, a two-room temporary classroom structure that the Pelham Board of Education had constructed to separate the PACT students from the mainstream school. After all, students sent to PACT were looked upon as dropouts, discipline problems, and dullards.

Al was a student who came to the alternative school for its "caring community." Al needed to improve his skills with operations of fractions. He had already completed his math requirements for graduation. Back in the 1970s students were able to use business math as a math credit. Al knew he was weak in math. He needed to learn the skills in order to construct the scale model of a house he was designing with Dr. Kasal's supervision.

Al came to school each day and worked on his math skills on a computer that generated a set of basic arithmetic problems (operations with fractions, decimals, and percents) and provided instant feedback for incorrect answers. These were math skills that Al could process, internalize, and apply to the building his scale-model house. The set of problems ranged from very easy to very difficult. Al's goal was to become fluent in calculations using fractions, decimals, and percents.

Al's goal was to achieve mastery (at least 85 percent correct on the quizzes). The computer scored his work and produced an answer sheet for Al with a model of how the correct answer was determined. If Al needed help understanding what he had missed, he asked the PACT math teacher. Most of the time, he was able to see that he had made a careless mistake. He repeated the set of quizzes on a particular skill until he reached mastery. He learned at his own pace because he knew he needed to be fluent in basic calculations to become a builder.

A twenty-first-century school needs to provide a clinical component where a student can self-assess and know what math skills and concepts he or she needs to work on. We see the student, thirty years ago, making sense of mathematics, applying math concepts and skills to a possible career as a builder. Al had the perfect mix of math connected to his interests. Math was meaningful and provided career and college readiness.

Al had immediate feedback and was able to set meaningful goals for learning more mathematics. He was able to scaffold from easy problems to more challenging ones. Twentieth-century instruction that embraced twenty-first-century technology, using computers to generate math problems and providing answers immediately, were involved in assessing Al's progress. Al had a teacher/facilitator available for help. He looked forward to working with Dr. Kasal twice a week. He also was able to meet with a group of other students in a structured math class.

Unfortunately, Al was still under the Carnegie Unit of credit system—he could not test out of any subjects or get a credit for his work. He did attend class with students of different ages and was not bound to a "cohort," attending class with everyone the same age; he received a credit for consumer math. PACT was a school of the future.

Maximo and his parents, on the other hand, are facing a foggy math instructional journey. There has been little feedback about what skills and concepts Maximo needs in order to improve. The good news is that Maximo has embraced the opportunity to meet with his math teacher after school for support. He has maintained a B+ average in seventh grade advanced math. The only feedback on his academic progress in learning mathematics will be quizzes and tests from math class.

Maximo attained only a level 2 on the state assessments for grades 3–5, but classroom assessments did not provide clarity on the exact skills and concepts Maximo needed to practice in order to achieve at least a level 3 on the state exam. Will Maximo's Common Core curriculum be switched to a new math curriculum as he enters high school?

To invoke math reform (shift from content- to standards-based instruction), there is a need to create an educational climate that is clinical in nature where a student's understanding and application of math concepts and skills can be evaluated, and where an individual instructional plan can be developed for each student. This will lead to schools where students become independent learners and collaborate on projects where mathematical concepts, skills, and critical thinking are applied to meaningful tasks.

Are there school districts that have shifted from the traditional content-based to standards-based instruction? Can failing school districts improve? How does the change take place?

Let's take a look at a successful school district—like one in Union City, New Jersey, where the graduation rate was 81 percent in 2014. Union

City opted for homegrown gradualism and is regarded as the model for good urban education ("How to Fix the Country's Failing Schools. And How Not To," by David L. Kirp, *New York Times Sunday Review*, January 9, 2016). Union City was able to develop a successful learning environment over time.

Kirp wrote how teachers rethought skill and drill instruction and instead emphasized hands-on learning and group projects. Help came in the form of coaches (veteran teachers) working side by side with newbies, where time was set aside for the teachers to collaborate. Students are frequently assessed, not to punish but to pinpoint areas where help is needed.

Bingo! Union City implemented the standards-based approach to instruction. Rigorous training for teachers, time for teachers to collaborate, veteran coaches, and frequent analysis of the students to "pinpoint areas where help is needed." Yes, there have been, are presently, and will be school districts that invoke change.

DO NOT FORGET THE IMPORTANCE OF CLASSROOM MANAGEMENT

It comes down to teacher training in classroom management. This needs to be a part of the student teaching practice. Body language is also a key to maintaining class decorum. Students are like barometers—they know when a teacher is vulnerable. Students can sense whether or not a teacher is prepared for class. Without classroom management techniques, instruction is sacrificed. Curriculum reform, shift, or engaging instruction cannot take place in a disruptive environment.

During student teaching, the cooperating teacher models classroom management techniques. Usually, the cooperating teacher allows the student teacher an opportunity to take over several classes. However, the student teacher gets a sense of what it is like to head up a class but not how to establish the behavioral climate in the classroom since that has already been established by the cooperating teacher.

When the new teacher enters their own practice, oftentimes he or she is not coached on how to manage a class and, by the time the teacher's classes are into the first week of the school year, lax discipline has taken root and cannot be corrected. The most important foundation a teacher needs in order to deliver instruction is the establishment of firm classroom management.

WHAT CAN PARENTS DO IN THE MEANTIME WHILE THE PENDULUM SWINGS?

While districts shift and change, what can parents do to ensure that their child gets the correct math instruction? There is no silver bullet that makes students "love" math. Feeling confident as a problem solver is key to a student's self-efficacy. Chapter 5 gave an overview of the Common Core math standards as they relate to your child's math program. Most of all, it is important that your child understands what it means to be a "math student." Learning is not always easy or fun but rather is the development of a passion to learn.

It is important that students understand what it takes to study math. Learning math is like playing a sport or becoming a musician. There is practice, there are rules, and there are strategies that need to be learned to solve problems. There can be both satisfaction and frustration in the problem-solving process. Students will learn more if the math is connected to their interests, and to college and career plans.

Focus on your child's self-efficacy (the extent or strength of one's belief in one's own ability to complete tasks and reach goals) rather than on a false sense of self-esteem. Children should be able to communicate what they know and what they need to learn. Self-esteem will improve when children believe that they have learned something they did not know, or improved in an area in which they were weak.

Check for your child's engagement in math at home, and how he or she relates to math in school. There is a "street math" that we all use that is not always like the math we learn in school. Street math is used daily and oftentimes we do not realize we are using math skills. When we figure out a bus schedule, time a pizza in the oven, play a video game, or read instructions and troubleshoot to find out why our digital devices are not working, we are living in *MathLand*.

What can you do as a parent to foster the love of learning math in your child? There are two paths that can lead to successfully supporting your child—learning to become proactive and not reactive, and learning how to advocate for your child in school.

PATH #1: BE PROACTIVE AND NOT REACTIVE

(A) Support your child's hobbies, interests, and passions. Math skills and concepts are connected to music (fractions), art (geometric shapes, proportions), sports (statistics and probability), technology (game simulations), science (equations, graphs), social studies (surveys), social media (learning online), Legos (modeling, seeing patterns), and entrepreneurship (financial mathematics).

It is through these interests that students learn to persevere and solve complex math problems. Perseverance is the first and foremost math practice. Math can be found in all aspects of life. Be aware that there is academic math and street math. Street math is the approach to problem solving in an informal setting, such as estimating on-the-spot when purchasing a product. Academic math is the formal math education that is needed for acceptance into college and vocational schools and for career preparation.

(B) Make sure that your child is fluent in: (1) number sense, including place value, multiplication facts, money skills—how to make change, and how to perform the operations of addition, subtraction, multiplication, and division with fractions, decimals, and percents; (2) identifying geometric shapes and solids; and (3) proportional reasoning.

Involve your child in measurement activities using rulers, measuring cups, and timers. It is essential for students to know how to read a ruler and any other measuring device (and to use the correct unit of measurement—inch, centimeter, liter, pound, ounce) and to convert units within the English and metric systems (feet to inches, centimeters to meters) as well as between systems (inches to centimeters, ounces to grams).

Following a recipe, learning music theory, perspective drawing, or following a favorite sports team's statistics are just some examples of how you can improve your child's number sense. Make it fun, a game, a contest.

Make sure that your child has reached the fluency in math required at each grade level. "Fluent" in the standards means "fast and accurate." The word "fluency" was used judiciously in the standards to mark the endpoints of progressions of learning that begin with solid underpinnings, and then pass upward through stages of growing maturity.

EngageNY is a guide as to what is a grade-appropriate fluency. "Grade K: Add/subtract within 5; Grade 1: Add/subtract within 10; Grade 2: Multiply/divide within 20 and add/subtract within 100; Grade 3: Multiply/divide within 100 and add/subtract within 100; Grade 4: Add/subtract within 1,000,000; Grade 5: Multi-digit multiplication; Grade 6: Multi-digit division and multi-digit decimal operations; Grade 7: Solve px + q = r, p(x + q) = r; Grade 8: Solve simple 2 x 2 systems by inspection."

In fact, the rarity of the word "fluency" itself might easily lead to it becoming invisible in the standards. Assessing fluency could remedy this, and at the same time allow data collection that could eventually shed light on whether the progressions toward fluency in the standards are realistic and appropriate (EngageNY).

(C) Do not do your child's math homework! It should be clear to you and your child why a homework assignment was given (refer to the introduction's homework map problem). The student is clear what the homework task is. Students need to be able to tackle independent work without the support of an adult outside of school. The support comes from the teacher fostering children to discuss and present solutions to math problems in the classroom.

Homework (independent work) needs to be assigned in order to provide practice of what was learned that day in the classroom as well as application of concepts and skills to real-world problems, tailored to each child. Students should leave the classroom with practice where results are available. They should know what they had difficulty in doing.

Parents, siblings, and friends can have homework discussions. Encourage your child to discuss different approaches to problem solving with peers and siblings. If that's the case, make sure the teacher knows that the work done for the next day was collaborative. One example of best practices that fosters independent work is the homework assignment that requires students to create a math problem, solve the problem, and then share the problem with others (viz., friends, parents, siblings).

(D) MAKE SURE THAT YOUR CHILD LEARNS TO READ and comprehends what he or she is reading. Many students know the math but cannot extract from the verbal math problem what the question is asking them to do, such as calculate or represent pictorially. Even if students know what to calculate, they have difficulty determining which variables to select to set up a problem.

An oft-overlooked training for math teachers is literacy in mathematics. You want to know how reading and writing mathematics is integrated into the curriculum. Ask how your child records notes from class that are used at home to decipher homework problems.

You want to know the reading strategies used by the teacher to help students comprehend the problems they need to solve. Ask if the teachers have been trained in integrating the use of two-tier (academic) vocabulary words and how to read a more complex text into the math instruction. Ask administrators if they use lexile (reading) and quantile (math) data to determine the grade level achieved by your child. This data will be helpful in providing remediation or acceleration math tasks to help develop your child's math acumen.

(E) What can be helpful is zeroing in on your child's math learning style as a preferred way to solve problems. Once students identify their preferred way to solve math problems, work on solving problems based on other styles. Remember, students' math profiles include all four learning styles.

For example, if your child has " Understanding" as a dominant style, provide opportunities for "Interpersonal" style group work, such as interaction in playing games with others.

(F) Encourage your child to take as many physical science (physics and chemistry) courses as he or she can include in their high school schedule. Students who study physical sciences become better math students. Why? Students apply math skills and concepts almost daily in physical science lessons. Physics is a gatekeeper course for college entrance. The more physical science and STEM projects that children engage in, the more adept they will become in their math skills.

(G) Do not buy into the idea that it is not necessary for students to study algebra. There are college professors who write articles regarding students needing to take algebra. The current idea is to have students avoid algebra and focus on proportional reasoning, a Common Core category (domain) relegated to elementary and middle school. While many careers rely on basic math skills where employees are not required to use algebraic equations, such an approach does not support students who apply to postsecondary schools.

Those who believe that students don't need to learn algebra fail to realize, however, that all applicants to community and four-year colleges must have at least three years of passing math grades in Algebra I, geometry, and Algebra II. Following the rationale of only being adept in proportional reasoning and not taking algebra, there would be very few students who would have qualifying transcripts needed for college applications.

(H) Take advantage of cyberspace. Students today are experts in the world of cyberspace. There are a lot of excellent opportunities to learn mathematics online.

It is important to learn how your child's school incorporates technology into instruction. Great questions to ask are, "Do all classrooms have Internet access and SmartBoards?" "Is there a math remediation website approved by the district?" And "Is the classroom a blended classroom or a flipped one?"

Are students allowed to use their own iPhones and iPads? Cyberspace is here to stay. Rather than dissuade your child from using his or her handheld devices, investigate how they can use them to study mathematics. There are teachers who have embraced twenty-first-century technology and encourage their students to use their electronic devices to enhance their learning. Help your child develop perseverance in learning online.

The important aspect of online instruction is the instant feedback that students get when they do a math problem. Online support websites, such as Castle Learning, offer practice problems for students K–12 in ELA and science as well as math.

(I) Be aware of the time spent on test prep versus time spent on instruction. Ask the teacher if they have had the opportunity to look at the statistics from their district that compare the 2015 test scores to the 2016 test scores, for example. Remember the statistics that point to the mathematically able students, where teachers use only 15 percent of their instructional time in review for the test, whereas the mathematically challenged students experience 47 percent of instructional time reviewing for the test questions. Does this really help these students?

There is a common misconception among educators that the more the questions that are reviewed resemble those on the test, the better students will perform. However, if students have not learned how to read the problem, identify what needs to be solved, and select the concept/skill to solve the problem, then the time spent on test preparation is not prudent.

(J) Feeling safe in school—is the school environment a safe one? Parents often remove their children from school where discipline is lax, then pay tuition to a neighboring school district where discipline is exemplary. When asked if the student left the school because of poor teachers, the general comment is that the teachers were good, but the discipline was lax.

A common scenario that occurs in schools where discipline is lax is when a fight breaks out in the hallway. A teacher who is teaching a lesson has to interrupt the lesson and either call the main office for help or try to intervene. Oftentimes, students do not get to class on time and tend to congregate with other students in the hall, where a fight inevitably ensues.

Classroom management is another form of lax discipline. If a teacher cannot control the class, no learning can take place. Elementary schools are usually calmer and have fewer discipline problems. Due to the nature of the classroom structure, students do not move from class to class. They cannot get out into the hallway. Students will usually encounter more discipline issues when they enter middle school since the nine-period day affords students lots of time to get into hallway incidents.

(K) Be aware (as a parent) of your math learning style.

- If "Mastery" is your dominant style, you may find yourself saying that there is no need for math reform; learning will occur with the math instruction delivered procedurally.
- If "Understanding" is dominant, you may not be sympathetic to group discussions and presenting arguments to support your approach to a solution to a problem.
- If you have a "Self-Expressive" dominant learning style, you may be turned off to learning math altogether.
- If you have an "Interpersonal" style, it may cloud your need for abstract reasoning—you want all practical applications.

Therefore, as a parent, your dominant style may prejudice your understanding of mathematics, with or without the Common Core. Use your learning style to help understand the importance of standards-based instruction.

(L) Be aware of mindsets and beliefs that students can't learn because of their socioeconomic background. The school culture might foster the attitude, "Why give homework if students can't do it?" preventing the assignment of online remediation websites to students because it is believed that their parents do not have the money to purchase a family computer.

PATH #2: BE AN ADVOCATE FOR YOUR CHILD IN SCHOOL

There is nothing wrong with advocating for your child in a constructive way, especially when you ask for data regarding your child's progress. Find out what skills and concepts your child knows as well as those that need improvement. As a parent, you can request state assessment information and district diagnostic exams from the teacher or principal.

The data can be used to create a baseline to show progress from where your child was last year. If their score was a level 3, it might be possible to prescribe what the child needs to do to fill the gaps and move up to a level 4 on this school year's assessment.

Parents' Night—Ask the principal to address how data gathered from state tests is shared with teachers and how it is used to inform students (to improve achievement). For example, the principal should discuss how the school performed on the state exams or normed assessments used to diagnose students' weaknesses and strengths in learning the standards.

Request that the teachers provide a scope and sequence of the math curriculum for their course. Do not accept the answer that, "We are using the Common Core Mathematics Curriculum"; do not accept that the math curriculum has been aligned with the Common Core. Ask for a copy of the math course outline or curriculum map. Sometimes districts post curriculum maps on the school website.

Often, teachers think that they are teaching the Common Core but they are still teaching traditional lessons. A good teacher will provide an example or strategy that they will be using in their instruction (such as place value charts to illustrate subtraction of whole numbers or decimals).

The district's counseling personnel should be able to explain if your child's level 2 is a low, middle, or high 2. Is the district able to identify the gaps in your child's understanding of math concepts and skills? The district should be able to divulge the plan for providing instruction for the child so that he or she can raise the score on the next state exam.

Remember that Maximo had achieved only a level 2 for the past three years (grades 3, 4, and 5) with no diagnosis or plan to fill the gaps. Asking for diagnostic data from the school is essential for planning remediation and enrichment that is meaningful to each student.

Note that school districts should not just rely on the state exam data. One of the issues is the lapse of time between June and September. Ten weeks is a long time. The summer erases any knowledge and understanding of math concepts that the student has not internalized. The lost knowledge and understanding may be retrieved to review or re-teach in the beginning of the year.

In his book, *Outliers: The Story of Success* (New York: Little, Brown and Company, 2008), Malcolm Gladwell identifies summer vacations as an educational vacuum that some families fill with experiences that give their children support that bridges understandings from June to September. Not all students are lucky to have that summer support.

There are alternate assessments that can be used in tandem with state assessments to identify student progress and the effectiveness of instruction during the year. Prudent school districts administer evaluative assessments that have been developed to appraise student progress with the Common Core standards periodically throughout the school year (for example, STAR, aimsweb).

These pre-tests are common and are used in many districts. Some use the STAR assessment to diagnose student strengths and weaknesses in both the Common Core math and ELA standards. The STAR assessment not only measures student progress during the year, but also has a series of reports that can be used to guide instruction by teachers and parents.

In particular, STAR provides a Student Individual Instruction Report that teachers can download for each student. The report identifies weaknesses of each student and provides a plan for which math standards need improvement.

Unfortunately, all too often, data (for example, STAR) is available, but teachers do not have the time or training to use it. Teachers may have been trained in how to get a general STAR report from the database, but they do not know how to access the individual student plan. To complicate matters, they can access the report, but they don't always have the time to make a remedial instruction plan to address each student's needs.

To ramp up the intensity of the perfect storm, districts may have STAR data but are not able to share it with teachers in a timely fashion to inform and plan instruction. As a result, parents are not being notified about their child's diagnosed weaknesses.

A teacher sharing the results of a child's state exam score taken six months prior in April 2016 at the October 2016 parent meeting is not informative or helpful; it just leads to frustration. Equally frustrating may be waiting until January 2017 for the STAR results for students who took the test in September 2016.

In a perfect world, each grade level, K–12, would have instituted a pre-test math program. The assessment programs, like STAR, generally allow a teacher to assess a student three to six times a year to look for areas that need improvement. Again, the structure of the school day, daily schedule, holidays, and snow days often interrupt the process of assessment. Implementation is often impacted by the time element—there is too little time to diagnose and no time to institute a remedial plan.

On the other end of the spectrum is the student who has achieved a passing level on the exam and needs enrichment. The question is, "Are we challenging the level 3s to go on to level 4, and the level 4s to do advanced work?" These students may be in a class that has many level 2 students and they are left to sit for lessons on concepts that they already know and understand. Why can't those higher-level students test out of the lessons they know and work on a more challenging curriculum?

Then there is the argument that all students need the "seat time," even if they can test out of a curriculum. Students won't get credit because they did not "sit" in the math class five times a week (180 days!)—the "Carnegie Unit" of the 1800s is holding back the advancement of more-able learners. STAR reports for the advanced students are off the charts, listing them at grade 13 or college level.

In conferences with their child's math teacher, parents often equate math scores with ability. If the scores are poor to mediocre, parents believe that math is not a strong subject for their child (and thus, phobias are born). The fear of learning mathematics is then carried throughout the child's school years. On the other hand, if the scores are tiptop, parents believe their child is a genius (a perception that may last through middle school and into high school), where math is taught procedurally, without engagement.

If Einstein and Newton were attending our schools today, they both would be placed in special education programs; Einstein would have questioned his math results, especially in his early grades, and Newton would not want to answer math questions when called upon because he would perseverate on why we can't divide by zero. Bill Gates and Steve Jobs are two documented examples of how our education system tends to lose the Einsteins and Newtons of today's world society.

We label students all the time—that they are not capable or are mediocre math students, or that they are super in math skills (graphing data) and facts (definition of a circle), but there is no indication of how well they can problem solve.

Then there is the myth of differentiated instruction, a noble idea, but very difficult to implement in classes of thirty-five students with a wide range of abilities. Consultants are often brought in to help teachers design differentiated lessons. Theoretically, a differentiated lesson provides challenging problems for all levels of student ability. Often, students who are high achievers welcome the challenge. However, creating differentiated lessons via a content perspective is time-consuming.

There is always an element of fear on the part of the teacher with regard to the rigor of the lesson. Simplifying math problems lends itself to a modified curriculum (for special needs students). There is more of an even playing field for students when lessons are differentiated by learning styles and they are more likely to "hook" students while still retaining the rigor.

Does your child's school provide such preventative instruction? Does the school support staff (guidance counselors, school psychologists, and social workers) assist teachers by incorporating test-anxiety reduction strategies that are age-appropriate?

NPR suggests the following to reduce test anxiety (National Professional Resources, Inc., Test Preparation: A Teacher's Guide (2010), www.NPRinc.com):

1. Reducing physical symptoms—a) Deep abdominal breathing can be easily taught and practiced; b) Relaxing muscles by first tensing different muscle groups; c) Students can perform simple stretching exercises in their seats (without distracting others) during a long exam.
2. Reducing emotional symptoms—a) Visualizing a scene that can transport students to a peaceful, calm setting; b) Meditation that focuses on breathing, or a mantra that can help students lower heart rates and blood pressure; c) Relating self-expression of fears to others.
3. Reducing mental/cognitive symptoms—a) Replacing negative statements with positive, but realistic, self-talk statements. Instead of saying, "I have never been good at math, so I know that I am going to fail the math test"; say, "Because math is not my best subject, I will ask my teacher for help and study extra hard." (NPR, 2010)

In schools of the future, students will come in, sit at their cubbies, and know what they need to work on. Students will be presented with a schedule for that day of any collaborative sessions they need to attend. There may even be a live lecture that students can attend. Students will attend school until 5 PM. They will complete their individual work (homework) before going home to be with their families.

The school year will be twelve months long. Students will be able to take trips with their parents at any time without being restricted to summer or holiday vacations. Holidays are for families! Even if families decide to go away for the summer (who has a vacation that lasts ten weeks?), students will be able to sign onto the school's website and be involved in either a live chat room, a lecture, or a webinar, and they will be able to work on concepts and skills during the time away from school.

Teachers will be highly trained, like doctors. They will be high-level clinicians who can evaluate students and prescribe appropriate learning tasks, thereby providing meaningful, independent work for students.

There will be neighborhood schools that will be networked, like hospitals, to provide diverse curricula. This model will prevail for elementary and middle schools. At the end of grade 8, students will be able to enroll in high schools that offer college- and career-ready curricula.

High schools will be modeled differently. Coming out of grade 8, each child will have a profile of learning strengths and gaps. There may still be traditional classrooms. However, there will not be just one teacher assigned to each course (or responsible for 150 students!), but rather a team of teachers, all of whom will be capable of delivering the lesson.

One of the teachers, the "lead" teacher and an expert in classroom instruction, will deliver the lesson. Another member of the team will be an expert in designing the lesson and will provide the lead teacher with all resources required for the lesson delivery. A third team member, the clinician, will evaluate the students and differentiate the tasks (to create appropriate independent learning tasks).

District infrastructure will differ based on the culture, socioeconomic status, and community resources available to "hook" students into applying math concepts and skills to solve twenty-first-century problems, but not without the help of parents advocating for rigorous instruction that engages students in all math instruction. *MathLand* can be navigated so that children experience success in their school math program and progress through their journey with a sense of pride, perseverance, and the belief that they can learn math.

www.ingramcontent.com/pod-product-compliance
Lightning Source LLC
Chambersburg PA
CBHW030113010526
44116CB00005B/236